Marketing like a Peacock in the Land of Penguins

A Practical Guide to Effective Marketing

Nadji Tehrani

authorHOUSE®

AuthorHouse™
1663 Liberty Drive
Bloomington, IN 47403
www.authorhouse.com
Phone: 1 (800) 839-8640

Published by AuthorHouse 03/28/2018

ISBN: 978-1-5462-2208-8 (sc)
ISBN: 978-1-5462-2207-1 (hc)
ISBN: 978-1-5462-2211-8 (e)

Library of Congress Control Number: 2017919501

Print information available on the last page.

This book is printed on acid-free paper.

Every Company Wants To Be A Peacock In The Land Of Penguins, But Few Companies Know How To Do It Right!

Contents

Dedication Page

To my dear wife Juliette, my son Richard and his wife Mariana, my daughter Michelle and her husband David, my grandchildren Priscilla, Nicole, Isabella and Dylan. Thank you for your support and loving kindness.

Acknowledgements

There have been numerous people who have contributed in the process of completing this book and I must express my gratitude. I offer thanks to each one of you for your support:

- Contents and contributions by Nadji Tehrani the founder of Telemarketing®, Call Centers and Teleservices industries

- Richard Tehrani, the contributor of the opening remarks.

  Richard played a key role in the development of TMC

- Robert Hashemian, Technology Advisor

- Michael Genaro, EVP of Marketing and Advisor

- Alan Urkawich, Art Director

- Michelle Tavdi, Editorial contributor

- Stephanie Thompson, Administration Director

- Paula Bernier, Editor

Foreword by Rich Tehrani

CEO of Technology Marketing Corporation
www.tmcnet.com

A Spectacular Journey Working With My Father Nadji Tehrani

For weeks, I wondered how to start this foreword. How can one condense a lifetime of experience into a page? Emotions, facts, milestones, experience – there is so much to share.

I was fortunate enough to be hired by my father to work summers, holidays and eventually after school in the mailroom at age nine. I made a full $17 per day which equated to about 4-5 hours in the neighborhood arcade. At ten, I had a bank account and small fortune while learning about business and honing my asteroid-shooting skills.

I've probably spent more time with Nadji than anyone. But how do I sum it up for readers embarking upon a word-filled journey into marketing and strategy?

Perhaps the best way is to start with a single word and work outward.

Visionary.

Yes, that sums it up. How could one person be so ahead of his time for so long I've often wondered?

Decades before there was a green movement, he launched a publishing company which focused on using additives to paints and coatings to reduce energy consumption by 98%. If you've ever seen glue or paint being cured by a UV light instead of heat, it's due to additives called photoinitiators he first educated glue producers about.

When he had to name this organization in 1972, he chose Technology Marketing Corporation (now TMC). Imagine picking a moniker which would endure for a half-century. Some of the hottest tech companies like Facebook and Google are in effect technology marketing corporations.

In 1982, he realized the phone was an incredible way to sell advertising and looked for a magazine to learn more. He couldn't find one and launched what was the first call center magazine in the world. Keep in mind this was about half-a-decade before the term call center even came to exist! At the time, using the phone in sales was called marketing by telephone or telemarketing. He described it as the magazine of electronic commerce and communications – once again, at least a decade-and-a-half ahead of the ecommerce and communications revolutions.

From there, he launched the world's first call center conference in 1986 in Atlanta, GA.

A decade later he saw the need for a publication in the VoIP market and as a result Internet Telephony Magazine was launched to cover what eventually became a multibillion dollar market transforming communications for consumers, carriers and enterprises.

At the time, we weren't sure we had the resources to launch this publication quickly enough to get into advertiser budgets but Nadji had taught us the importance of being first. We worked day and night to launch quickly and the publication has been in production for two decades and has a sister show, ITEXPO which continues to thrive.

His expertise in marketing was a helpful guide throughout these journeys.

Being able to work with a family member for a lifetime and still talking with one another is an accomplishment to be sure. Like any family/business relationship it can be challenging but also rewarding. We witnessed the call center market evolve from using index card databases to mainframes,

UNIX, PCs, servers and the cloud. We saw communications literally collide with computing and networking, not to mention cloud and internet.

All of this was happening while our own market, media was going through numerous transformations.

Throughout this tumultuous environment where you had to innovate or die, we were fortunate to work with Nadji and his advanced vision and eternal optimism. As I finish this foreword I wasn't even sure how to start, I hope a little bit of him rubbed off on me.

When you finish this book, you'll hopefully say the same.

Introduction

"La Raison D'Etre" – Marketing is Paramount for Business Profitability and Success

Positioning and Differentiation: Give Your Customers Reasons to Buy From YOU!

The most important department in any company is the marketing department, PERIOD!

This is "la raison d'etre", the purpose, for writing this book. Marketing must be first and foremost in every organization. You can create any product as good as gold, but without proper marketing, no one will buy it. Marketing is the No. 1 source of success for all companies, new or old.

According to a former professor of marketing at MIT, companies exist for two reasons:

1 - Marketing
2 - Innovation

If you are lacking in either one of these, YOU WILL NOT SUCCEED.

On Differentiation And Positioning: Every Company Wants To Be A Peacock In The Land Of Penguins, But Few Companies Know How To Do It Right!

As founder of Technology Marketing Corporation, Nadji Tehrani has written extensively for the past 35 years on marketing strategy. He has been recognized as one of the most influential thought leaders in the world.

Nadji has met a number of people who have asked him why he stopped writing his greatly anticipated monthly editorials in his media company's many publications.

He responded, "At some point I needed to step aside and my son, Rich Tehrani, has done a terrific job running the company and providing insight to our many readers and followers."

In this book you will find valuable and easy to follow guidelines which when implemented properly will propel your marketing campaigns to unparalleled success.

The Missing Link in Marketing: Differentiation and Positioning

Your customers must have a reason to buy from you and that reasoning comes from positioning and differentiation.

Definitions

In order to better understand the purpose of positioning and differentiation, which are the most crucial parts of marketing strategy, we thought it would be a good idea to refer to *The American Heritage Dictionary* and find the definitions offered for the words, "differentiation" and "positioning." Although the dictionary does not provide a direct definition for differentiation and positioning in marketing, if you look at the definitions for "differentiate" and "position," you will arrive at the same conclusion.

American Heritage defines "differentiate" as follows:

- to constitute the distinction between
- to perceive or show difference in or between; and discriminate

"Position" is described as follows:

- the right or appropriate place
- the way in which something or someone is placed
- the act or process of positioning
- to place in proper position
- an advantageous place or location

Based on these definitions, you can clearly conclude that to effectively market, any product or service must be differentiated from its competition, giving the potential buyer a reason to purchase.

As for positioning, the definition implies it is crucial for any product to be positioned in an appropriate place or, preferably, an advantageous location. If you don't position yourself advantageously, your competition will position you and your product in the most disadvantageous way.

Positioning and Differentiation, Like Marketing itself, are Not Part-Time Jobs

In fact, to do it right, they are more than full-time jobs. That means you must market *every day*, you must position *every day* and you must differentiate *every day*, 365 days a year, 24 hours a day, 7 days a week. In short, marketing, positioning and differentiation are full-time, 24/7 jobs, period, end of story.

Positioning Must Be Adjusted to Rapidly Changing Market Conditions

In today's ultra-fast-moving and rapidly changing environment, you can practically assume that market conditions also change month to month, maybe even day to day as opposed to 25 years ago when things changed more slowly. Consequently, you must always remain 100 percent focused on the marketplace as well as on the validity of positioning vis-a-vis the current conditions of the marketplace.

It would be a disaster to lose sight of adjusting your positioning to reflect the changing marketplace requirements. When companies fail to change their positioning, they lose market share and ultimately lose considerable sales revenue. In fact, these companies may not even survive when markets change so rapidly.

What Then Is The Proper Course Of Action? REPOSITIONING

Companies must remain extremely flexible and change as rapidly as the marketplace changes. This process is called, *Repositioning*, which is crucial to the survival of any successful and progressive organization.

How To Cope with Information Explosion

In today's extremely complex, information-jammed world, we are exposed to thousands of advertisements, promotions of various kinds and, in short, are inundated with information explosion. It has been said that in the last 30 years, more information has been produced than in the previous 5,000 years. The emergence of the Internet has added ultrasonic speed to the growth of information available.

Therefore, to make your products and services stand out in the marketplace, you must do a superb job of positioning, differentiation, marketing and advertising.

You Don't Want To Be a Penguin, You Want To Position Yourself as a Peacock

Take a look at the picture above. What stands out? Obviously, the peacock.

What sets the peacock apart? The magnificent colorful feathers and its artistic design vs. the bland black-and-white feathers of the penguin.

If you were to buy one of them, which one would you buy, the peacock or the penguin?

Those who would buy the penguin would need to see a psychiatrist!

The bottom line is, the peacock is different from the rest of the crowd; it stands above the rest with magnificent and attractive colors in the bland land of the penguins.

If you truly want success in your positioning or the position of your company or product, you don't want to be a me-too or a penguin. You want to be unique and position yourself as such. In short, you want to position yourself as a peacock in the land of penguins.

That is how you gain market share, penetrate the minds of the buyers and become a leader.

Also, you want to be FIRST, because nobody cares for second place.

The First Law of Marketing: Positioning

"It Is Better To Be First than To Be Better"
Position your company to stand out from your competitors!

Who was the first man who flew over the Atlantic?
The answer: Obviously, Charles Lindbergh.

Who was the second person to fly over the Atlantic?
The answer: Nobody knows and nobody cares about No. 2.

What was the name of the horse that won the Triple Crown in 1973 and broke practically all racetrack speed records?
The answer: Secretariat.

What was the name of the horse that came in the No. 2 position right behind Secretariat in all three races?
The answer: No one remembers and no one cares about No. 2. (Only horse racing fans would remember the name of the second-place horse, which was Sham.)

The bottom line: The first law is true and if you really want to be a market leader, you must have a strategy to position your business every minute, every hour, every day, every month, 365 days a year, 24/7.

The Second Law of Marketing:

"If You Are Unable To Be FIRST In Your Field, Then Find A Niche In Which You Can Be No. 1"

The validity of the second law is also justified by the reasons given for the first law.

Finally, don't forget "Repositioning" as markets continue to change!

Differentiation Makes THE Difference

Differentiation is by far the most critical function in a successful marketing organization. The primary reason is that differentiation will give buyers a compelling reason to buy your product.

Without differentiation, no one has any urgency to make the purchase with you. For a thorough understanding of how to differentiate your company from your competitors, keep reading or jump to *Chapter 1: Differentiation – You Must Be No. 1, Nobody Gives a Damn for No. 2; How to Position Your Company as the Industry Leader.*

The Rules for Success

Chapter 2: The Rules for Successful Marketing in Today's Online World lays out the roadmap for navigating your marketing journey. Whether you have years of marketing experience or are just getting started, the very specific guidelines provided will show you the way to success.

Qualified Sales Leads are the Lifeblood of Every Company

Generating *qualified* sales leads is necessary to grow your business.

Many companies though, simply do not have sufficient knowledge or expertise to be successful in identifying the most qualified potential new customers cost effectively. We will teach you 12 guidelines for developing effective marketing strategies to create qualified sales leads in *Chapter 3: How to Create **Qualified** Sales Leads.*

"One for the Thumb" – The Five Championship Rings

To use a popular sports analogy, nineteen different American football franchises, including teams that relocated to another city, have won the

Super Bowl. The Pittsburgh Steelers (6–2) have won the most Super Bowls with six championships, while both the San Francisco 49ers (5–1) and Dallas Cowboys (5–3) each have five wins.

"One for the Thumb" was the rallying cry of the Steelers back in 1981 as the aging core of a four-time Super Bowl championship squad made one last push for a fifth ring. Pittsburgh fell short that season, finishing 8-8 and failing to qualify for the playoffs. It marked the beginning of what was for them a relative dry patch that would last until the 1992 season.

Since then the Steelers have qualified for the playoffs 13 times in 19 seasons. They also earned that "One for the Thumb" with a win in Super Bowl XL as well as another for the ring finger on their other hand with the Super Bowl XLIII title.

American football legend Charles Haley played in the NFL from 1986 to 1999. He has won the most Super Bowls of any other player with five in total, making him the only individual player to be able to wear all five rings, one on each finger, along with "One for the Thumb". Notable players who have won four Super Bowl rings are Joe Montana, Bill Romanowski, Adam Vinatieri, Ronnie Lott, Matt Millen and Terry Bradshaw.

So what does this have to do with your business? We have identified the five winning marketing strategies that will position your company as the champion of your industry. Because these principles are so important, a separate chapter has been dedicated to each of the five concepts:

- Award Marketing – Chapter 4
- Trade Show Marketing – Chapter 5
- Permission Marketing – Chapter 6
- Loyalty Marketing – Chapter 7
- Social Marketing – Chapter 8

Mastering all five of these strategies will give your business membership to the exclusive, "One for the Thumb" club.

Information is Power and Leads to Competitive Advantage

As the old saying goes, "he who holds the gold, makes the rules." Well the gold today is mined through data analytics. Marketing intelligence is the everyday information relevant to a company's markets, gathered and analyzed specifically for the purpose of accurate and confident decision-making. To understand the process of tapping into the information available to marketers for enhancing strategy, refer to *Chapter 4: Award Marketing – Winning With Strategic Competitive Intelligence.*

The Show Must Go On...

Effectively preparing and exhibiting at industry trade shows is a skill that must be developed by B2B marketers. This key area can be a significant source of new qualified sales leads for your company.

Don't even attempt to exhibit at an industry trade show until you've read *Chapter 5: Trade Show Marketing – The Basics To Be Effective.*

How Do You Gain Access to Decision-Makers: It's Easy – Ask for Permission

A real problem today is that B2B salespeople rarely ask for permission. You would likely agree when we say that less than 1 percent of sales pitches begin with the most basic courtesy of asking the customer for permission to present an offer.

If your company is struggling with opening doors to potential business opportunities, proceed to *Chapter 6: Permission Marketing – The Secret to Engage High Level Decision-Makers.* We share the secrets to engaging with gatekeepers and gaining favor with the ultimate decision-makers, which will propel you to greater sales success.

The Era of the Customer

Chapter 7: Loyalty Marketing – Customer Retention Depends upon Customer Care™ will change the way you look at your customers. We have clearly entered "The Era of the Customer."

With intensifying competition and ever-increasing customer demands, a company will live or die by the level of Customer Care™ and service it provides.

Providing relatable Customer Care™ requires being adaptable, similar to a chameleon changing colors to blend with its environment. Listen carefully to the tone and mood of each customer, and use a similar individualized approach to develop rapport with the person. Customer Care™ must be flexible, and there is no one method or approach that works in every situation. Recognize that everyone you come into contact with possesses unique personality traits. By attempting to understand them and adapt to their personal circumstances, they will be more likely to respond positively to you.

The perfect companion book to read on this subject is,
Taking Your Customer Care to the Next Level™
by Nadji Tehrani, along with Steve Brubaker

Saving Face and Avoiding the Twitterstorm

We all know social media has become the go-to method for people to share the good, the bad and the ugly of customer experiences. What if you could take advantage of this powerful medium to engage more closely with your customers?

Why not start today?

Chapter 8: The Do's and Don'ts of Social Marketing – From Followers to Brand Ambassadors, provides a step-by-step guide to the most effective ways to develop a legion of followers who are likely to share your powerful brand messages with their friends, family and other social network connections. Word of mouth has always been critically important to brand marketers.

Social media helps you take your message to a whole new level, if and only if, you know how to do it. We will teach you about the successes of others, as well as show you how to avoid the most common mistakes.

Employ the Right People

Hiring is a major problem in corporate America. That is why the tenure of marketing staff is extremely low.

Have you ever wondered why the chief marketing officer is short-lived in most organizations?

The average tenure of a CMO is just 45 months, according to a 2014 study released by executive recruiting firm Spencer Stuart. That's nearly double where the average tenure was as recently as 2006, when it was 23 months, but still only half the average time a CEO survives in most companies. In essence, CMOs are being driven out of organizations well before they are able to successfully implement a consistent customer-centric strategy.

Suzanne Vranica of The Wall Street Journal interviewed Greg Welch, global consumer goods and services practice leader at Spencer Stuart. Welch said: "Corporations expect chief marketing officers to prove their work has a real impact on driving the business. The CMO chair is a hot seat."

The WSJ reports that although CMO tenure has recently increased, marketing executives are feeling even more pressure to produce immediate results. Companies now "have different expectations of what marketing can do," said John Hayes, CMO of American Express. "There is an expectation that it's more measurable, more targeted and therefore more effective. This has made everybody's expectations higher," he added.

Marketers must be comfortable communicating with customers seamlessly across multiple channels and effectively deliver content across a myriad different devices from tablets to smartphones. Big data has produced a tsunami-like effect on businesses as well, meaning that the rules have changed forever, and today's CMO must evolve or risk becoming extinct. "Unless the CMO demonstrates the value that marketing is delivering and clearly shows the return on investment, the business owner, who is under pressure to deliver numbers, will pull funding," said Raja Rajamannar, MasterCard Worldwide's CMO. "The CMO could find himself out of the job."

Two basic factors are contributing to CMO churn. First, as we've already stated, those CMOs who do not produce immediate ROI are promptly put out to pasture. And secondly, when a CMO does deliver results, the world is watching and offers pour in from competitors and other businesses.

In *Chapter 9: Hire for Marketing Aptitude, Not Just Experience – The 44 Characteristics of the Right People*, you will learn the basic questions to ask when interviewing, which will quickly identify innate marketing talent. Then we will show you how to develop your team members' skills to take on your competitors with a winning strategy.

For example, consider a farmer whose objective is to utilize an animal that could climb a tree. It is clear he would not choose to use a pig or cow. Of course not! The farmer would get a cat or a squirrel instead to climb the tree, as those animals are better suited to the task. It sounds juvenile but makes sense, doesn't it?

This same basic rationale applies to the search for a qualified marketing employee. If you hire staff without a clear understanding of the importance of marketing, they will not be successful and neither will your company. The same goes for sales.... Don't be surprised, but we have found that more than 90 percent of people, even those with a business marketing degree, are unable to properly define the function and purpose of marketing in a concise way. Far too many marketing teams today are running in circles doing what they have been taught – trying only to create awareness for their companies' products and services.

When interviewing a candidate, right up front, ask him or her to define marketing and to define sales. In our judgment if a candidate cannot answer these two simple questions properly, there is no point in hiring the person.

We will provide you with our proprietary marketing aptitude test in *Chapter 10: Introducing "The Test For Marketing Ability™" To Help You Hire Talented Marketers – Our Proprietary Solution Published for the First Time Ever.*

The test begins with two basic questions:

Question #1: Define Marketing
Question #2: Define Sales

Actually, the rest of the marketing test is far more challenging. There is no need to proceed any further with an interview if the person does not know the answer to the first two questions. Here are the answers:

> Question #1: Define Marketing
> Answer: Generate <u>Qualified</u> Sales Leads
> Question #2: Define Sales
> Answer: Take the <u>Qualified</u> Sales Leads Generated from Marketing and Convert them to Customers

This book provides answers to all of the questions of our marketing test and will teach you to hire staff with proper marketing knowledge.

Marketing and Quality Are NOT Part-Time Jobs

Obviously in everything that we have stated to this point, we have communicated that without marketing no business could exist long term, and without quality product and superior Customer Care™, the customers will not return.

Henry Ford once commented that his customers were able to choose a car in any color, *as long as it was black*. He had no interest in providing customers preference in their selection. When competing manufacturers began to offer multiple color choices, Ford's marketing share began to erode. The lesson is you must listen to customers carefully to determine what they want and how to please them. The customer is always right!

Under Promise and Over Deliver

One of the commonly made mistakes among salespeople is to overpromise and under deliver. Of course, this method of selling is the most damaging thing any sales department can do. Customers are intelligent and have long memories. Once they recognize their experience does not match what they were told, they are not going to return to do more business with that company.

Testimonials and Third-Party Validation: Customer Ambassadors

Your current customers will be extremely effective in sharing your expertise with new prospects and should be a strategic part of your sales efforts. There is no better way to introduce customers to your business and reinforce your sales messaging than to have a currently engaged and satisfied customer tell the story of his or her success in partnership with you.

Customer testimonials provide third-party validation that is impossible to ignore. In fact, they are the most effective way to position your product and company.

Don Palmer, founder and former CEO of SIP Print, was so happy with his advertising and the ROI on his investment with TMC's CUSTOMER™

magazine, he was willing to call any prospects who were not familiar with TMC and tell them as follows:

"We built our company strictly by advertising in CUSTOMER™ magazine."

Don was further available to receive phone calls from prospects who wanted to advertise with TMC but needed more familiarity with the process. After one call with Don, customers would call back and commit to spending resources on advertising campaigns. That type of third-party validation is the most effective testimonial, where your customer is 100 percent dedicated to help your company grow.

Don was a regular advertiser and exhibitor until he sold his company. If you can manage to find customers as good as Don Palmer, by all means treat them like gold as they can be a major contributor to your company's growth.

Chapter 1

Differentiation – You Must be No. 1 Nobody Gives a Damn for No. 2; How to Position Your Company as the Industry Leader

Differentiation Is the Key for the Successful Marketer

Marketing does not begin with a great idea or unique product. It begins with customers – those people who want or need your product and will actually buy it. They aren't just buying your product/service, however. They're also buying the concept of what the product will do for them, or help them do for themselves.

Find Your Niche

One way to create a customer base is to focus on specific groups of potential customers that share common characteristics. Your product/service can't be everything to everyone. Rather than try to fulfill an illusory/impossible demand, spend time researching the market to find out who is most likely to buy your offering and sell to them. Not only is this common sense, but you build in-depth knowledge of their profile, as well as the industry, and you develop an environment of trust, which translates into customer loyalty and, thus, customer retention.

Once you've evaluated your market, identify your competition. You can't position yourself or your product unless you know what your competition is doing. Don't underestimate them under any circumstances. Try to counter their underlying strategy, not necessarily their exact movements.

Don't get caught in the game of one-upmanship, it will make your company look foolish and provide plenty of fodder for the competition to use against you.

Be aware that there is, by definition, a cap to your niche. Take early steps to streamline your organization so as to garner the highest profit you can while still providing the highest level of quality service and product to your customer. If you obtain new customers but can't retain them, your company will fail. Also consider the development of multiple niches with new lines of products/services related to the original. This strategy minimizes capital expenditure, thus enabling you to access related niches with a minimum of effort.

How to Position Your Product/Service

As we stated earlier, positioning is your competitive strategy. The following questions should help you define what you need to be doing, provided you're not already doing them.

- What do you do better, faster, cheaper, with higher quality that differentiates you from your competitors? Don't be vague; be specific.
- What is your target market? Does it know what makes you different?
- How are you letting the marketplace know?
- What techniques do you use to make the point?
- Are your advertisements and public relations efforts spreading the word you want spread?
- Is your ad agency or PR firm (these could be internal departments) aware of your unique solution, and are they making every effort to promulgate that difference in every form of media they create?
- How are you driving the point home at trade shows?
- Do you conduct pre-, during- and post-trade-show marketing? (If not, you should be.) If so, are you stressing your unique qualities?
- Are you using database marketing to contact those with whom you have an existing relationship in an effort to cement that

relationship? Are they aware that product/service to which they are accustomed can only be found with your company?

Differentiation Points

You must carefully select the ways in which you will distinguish your company's products/services from the competition. The differences you advertise should be:

- *Distinctive* – either isn't offered by others or is offered in a more distinctive way by your company
- *Important* – delivers a highly valued benefit to a sufficient number of buyers
- *Superior* – is superior to other ways to obtain the same benefit
- *Communicable* – is communicable and visible to buyers
- *Preemptive* – cannot be easily copied by competitors
- *Affordable* – the buyer can afford to pay for the difference
- *Profitable* – the company will find it profitable to introduce the difference

Your advertising should be bold and go against the grain. Don't be offensive, by any means, but do be original. Every form of media generated by your company or its proxies should continually and eloquently emphasize why your product is the one of choice.

Hazardous Waste

As in all things, there will be copycats who will make every attempt to supplant you from your niche by mimicking your actions and positioning themselves as leaders. And while they add little value to the marketplace, they do perform a vital function – they make you look good. They force you to maintain your competitive edge.

That is why monopolies seldom live up to their potential – there is no outside presence to keep those companies on their toes. As long as you continually position yourself as a leader and you can back up this assertion with verifiable results and achievements, your position will remain secure.

But do not forget, ever, that perception is reality. If you don't create an image for yourself, your competition "never overlooks a mistake, or makes the smallest allowance for ignorance."

If Business Gets Any Worse We Should Probably Start Advertising

As lifelong students of marketing, we have always been amazed that so many marketing managers, directors and executives just don't follow the basic principles of the discipline, which is unfortunate for both the companies they serve and their own careers.

Marketing is the lifeblood of every corporation. As we've stated: "Companies exist for two reasons, marketing and innovation." As strange as this statement may seem (to the people who have no clue about marketing), marketing is, in fact, even more important to the survival and prosperity of any company than stated above.

1. If you don't market, you don't exist.
2. Marketing is not a part-time job.
3. Unlike the common belief of most high-tech companies, marketing should not be regarded as a necessary evil, but rather it should really be the other way around!

No company can go very far or even exist without a well-prepared, strategically sound and realistic (cost-effective) marketing program.

The lackadaisical approach of high-tech companies toward marketing explains their high failure rate, a rate that is, in fact, higher than any other sector of business.

The Case for More Aggressive Advertising in a Slowing Economy

The worst thing any advertising executive can do is to stop advertising in a slowing economy.

Here is what an article in the Harvard Business Review stated:

"Advertising in an economic downturn should be regarded NOT as a drain on profits, but as a contributor to profits."

It is a known fact that all sales begin with qualified sales leads. The majority of sales leads come from effective and regular advertising. In short, there can be no new sales without new sales leads. If the economy slows, naturally sales and revenues of most, if not all, companies decrease, and about the only way to compensate for that is to bring in new business. And new business comes from new qualified sales leads generated from continuous and effective advertising. If anyone has any problem understanding this basic and elementary fact, that person should not be in business.

Advertising Is Vital to Reverse the Loss of Business

During a good economy, one might lose about 30 to 50 percent of its business due to natural attrition. In a slowing economy, this number is more like 50 to 75 percent of business lost due to natural attrition. It is vital to replace this attrition with new business, meaning new dollars coming from new sales leads and continuous advertising for new products and services offered by any company. It is really just as simple as that.

A Review of Positioning Laws

One of the most informative articles we ever read about positioning stated that the great marketing battles are not fought in the business environment, but rather in the minds of the consumers. After all, the ultimate essence of advertising, marketing, PR, etc., is to influence consumers to prefer and purchase the brand of products or services that are being advertised. Therefore, the most important by-product of advertising, positioning and marketing should be to win over the mind share of consumers or, in the case of business-to-business transactions, the mind share of end users.

An article in San Diego Executive magazine stated:

"If during an economic downturn you maintain a strong advertising presence while your competitor cuts his budget, you will automatically increase your "Share of Mind."

In a related article, ABM (American Business Media) stated the need for advertising in a different way:

"When times are good, you **SHOULD** advertise; when times are bad, you **MUST** advertise."

Here is another comment regarding this subject matter from Coopers & Lybrand:

"During an economic downturn, a strong advertising/marketing effort enables a firm to solidify its customer base, take business away from less aggressive competitors, and position itself for future growth during the recovery."

You Can't Argue With Success

"History has proven companies that maintain or increase their advertising investments in periods of economic downturns increase their sales and share of market, both during and after the downturn."

Source: American Business Media.

The Best Way to Increase Your Market Share

As indicated above, most marketing managers and executives tend to do the wrong thing during an economic slowdown. In other words, instead of aggressively advertising and marketing, they do just the reverse, and that usually spells disaster for their companies. In fairness, however, one cannot always blame the marketing or advertising managers of companies for doing absolutely the wrong thing by cutting the advertising and marketing budget of a company.

The necessity for aggressive advertising in a slowing economy must be understood by everyone in any company, particularly by the CEO, CFO, as well as by sales, marketing and advertising senior management. While the marketing manager might remind the CEO that the company simply cannot stop advertising in these periods, he or she is often overruled by someone else who has no clue about the vitally important necessity of advertising in a slowing economy.

On the other hand, this unfortunate elimination of advertising budgets by the majority of companies can have a positive effect in terms of market share swing. In other words, the ill-advised decision by 80 to 90 percent of companies not to advertise creates a tremendous opportunity for savvy companies to redouble their advertising efforts during a slowing economy, thereby significantly increasing their market share while their competition is napping. History is full of examples of companies that lost their market leadership due to lack of marketing during a slowing economy.

Marketing Is Not a Part-Time Job

Everyone in the entire organization must understand this vitally important, basic principle of marketing, and unless everyone supports it, the company will end up going nowhere. It is not enough to periodically advertise when the CFO tells you there is money available to advertise.

It is not only imperative to market and advertise on a regular basis, but your advertisement must also be powerful and highly effective.

To make an ad memorable and highly effective, often you need to go against the grain. If your competition is using Method A for advertising, you definitely want to use Method B, which is totally different from Method A, to thereby make your products and services stand out.

Business-to-business ads, particularly in the case of niche-type businesses, that are highly focused and placed online and in targeted magazines are extremely effective, provided they are well prepared and response driven.

Getting back to the need for advertising in a slowing economy, here are some more excerpts from American Business Media (ABM) on this topic:

1. **"Maintaining or increasing advertising budget levels during economic downturns may be necessary in terms of protecting market position vis-a-vis forward looking competitors."**

2. **"If a company fails to maintain its 'Share of Mind' during an economic downturn, current and future sales are jeopardized. Maintaining 'Share of Mind' costs much less than rebuilding it later on."**
3. **"Advertising through both boom and down times sustains the necessary brand recognition."**
4. **"Maintaining a company's advertising during an economic downturn will give the image of corporate stability within a chaotic business environment, and give the advertiser the chance to dominate the advertising media."**
5. **In addition, The Strategic Planning Institute states: "Economic downturns reward the aggressive advertiser and penalize the timid one."**

In Summary

From all of the above, one should conclude that as the economy slows, advertising not only must continue, but also must increase because this is the time when market share changes hands. Second, it is the job of marketing, sales and advertising managers to inform senior management that it is vital not to cut the advertising budgets during the slowing economy and, in fact, the budgets should be doubled to be effective and to maintain and indeed increase market share.

Niche Marketing

The First Law of Marketing: It is better to be FIRST than to be BETTER in marketing.

No matter how good your product is, without marketing you cannot sell much.

On the other hand, if you made a good product and supported it with powerful integrated marketing, then you have a good chance of being successful. Follow the guidelines provided for effective digital marketing outlined in *Chapter 2: The Rules for Successful Marketing in Today's Online World.*

Remember that online marketing will not succeed unless and until you read and understand the guidelines provided in this book. Then and only then can you expect to see results from your digital marketing.

In the 1960s, when computing began to emerge as a solution for business processing, the technology was basically controlled by IBM. As other competitors emerged in the market, they were unable to outsell IBM, as even their systems had to be IBM compatible. IBM had developed the core mainframe computers and complex PBX phone systems for most organizations, and it was successful in locking competitors out due to compatibility requirements. Customers were contractually bound to long-term arrangements in which they were essentially forced to upgrade to the latest hardware.

IBM was first to market and it was able to quickly position itself as the leader. Big Blue was No. 1, and as a result took the majority of market share for many years. IBM's dominance persisted until personal computers became an alternative. With competition, other products became available which were often more powerful, but always offered at a lower price.

Apple has performed a similar feat in today's environment by maintaining a somewhat locked environment to its operating systems. Microsoft, on the other hand, provided flexibility for developers and others to connect to its systems and offer complementary hardware and software.

Nadji remembers his son, Rich, ordering computer parts and putting them together for a fraction of the cost of purchasing a pre-built system.

The first law of positioning is worth repeating: It is better to be FIRST than to be better.

The second law of marketing: if you cannot find a way to be FIRST, it is better to find something else where you can be FIRST and go to market that way.

You must be No. 1! We mentioned this earlier, but we will review it again here with added emphasis.

Who was the first person to fly solo non-stop over the Atlantic from the U.S. to Europe?
Charles Lindbergh.

Who was the second?
You can't recall, can you? That's because, "No one cares, no one gives a damn for No. 2." You must be No. 1!

Let us give you another example:
In 1973, which horse won the Triple Crown of Thoroughbred Racing in the U.S.? Secretariat, of course!

Which horse was second that year?
You can't remember, right? That's because, "No one cares, no one gives a damn for No. 2." You must be No. 1!

In the longest distance race in 1973, The Belmont Stakes, Secretariat came in ahead of the next horse by thirty-one horse lengths in a still record time of 2:24. The second horse, actually named Sham, came in last that day, even though Sham had been a close second to Secretariat in both the Derby and Preakness that year. No one really remembers Sham, even though the horse was very good and came within two and a half lengths of Secretariat in the Kentucky Derby that year, and Sham actually beat Secretariat in a couple of the Derby prep races that same year. Again, No. 1, Secretariat is the horse history remembers!

Last year, there was a new Triple Crown winner: American Pharaoh. American Pharaoh not only won the American Triple Crown in 2015, but also won the Breeders' Cup Classic the same year, to become the first horse in history to have won the "American Grand Slam" of horse racing. However, Secretariat would have dominated American Pharaoh at the Belmont Stakes as his 1973 time is still two full seconds shorter.

Don't waste your money by going out in second position.

The Art of Positioning: It All Begins With Effective Advertising. But Then: Why 90 Percent of Ad Dollars Are Wasted

Advertising professionals know that the heart of any campaign is the product and the position it holds in people's minds, and what helps to position the product or service in the consumer's mind is effective advertising. Now the proverbial $64,000 question is: Just exactly how many companies are using effective advertising? Would you believe less than 10 percent?

The Diagnostics Of Non-Performing Ads

The complexities of developing effective advertising are usually underestimated, which is, in our opinion, the No. 1 cause of failure for most ads. Here is a checklist of some of the important factors that contribute to the development of superior or inferior advertisements:

1. *Headline*

It is only common sense to believe that all ads must have a powerful, benefit-driven, highly creative and innovative headline that literally creates a major desire on the part of the reader to want to read the ad.

Without a powerful, benefit-driven headline, the likelihood of reading the rest of the ad is very close to nil. It stands to reason that decision-makers have very little time to waste trying to read every page or every ad without knowing what they will get out of the ad in advance.

2. *Ad Copy*

"The first 10 words of copy in an advertisement are more important than the next 10,000 words!"

That is the cardinal rule of integrated marketing – never forget it!

After the headline, perhaps the next most important ingredient for a successful ad is the copy. Copywriting is a highly specific talent, which takes years to develop. Unfortunately, this is not very well understood and most small companies (particularly, those led by entrepreneurs) throw a few words together without knowing anything about the highly sophisticated and complex factors that go into making an ad readable or desirable to read.

One of the greatest problems with copywriters is that by nature they like to write wall-to-wall text in an ad. Long ago, that concept may have worked in direct marketing, but today it simply does not work. In today's business environment, each one of us is exposed to hundreds, even thousands, of ads per week and, therefore, unless the ad is truly unique, creative and innovative with a great deal to offer in a brief statement, readers may pass over your ad rather than trying to read wall-to-wall text and waste their precious time.

3. ***Obscurity***

Many ads simply are the triumph of style over substance and not only are they not benefit driven, but they also do not tell the reader what it is that they are trying to sell. Such ads usually look good, but because they have nothing to offer, most people don't even bother reading them.

4. ***Creativity***

We have already established that every ad is competing with dozens and dozens of other ads for the attention of the reader. If your ad is truly spectacular and attention grabbing and would say in less than a few seconds what the reader will get from reading it, then that ad has a good chance at success.

5. ***Innovative and Outrageous***

Indeed, an ad that is highly innovative and outrageous will not only draw the attention of the reader but, also, it will remain in the mind of the reader for a long time.

One such ad appeared in a yacht magazine. While most yacht manufacturers were using young ladies in bathing suits in their ads to promote their yachts, one company chose to use an extremely old lady, perhaps in excess of 100 years old, sitting on the yacht and saying,

"When you have the best yacht in the market sonny, you don't need sex to sell it."

That ad clearly stood out and was successful because it was outrageous, innovative and it went against the grain. In other words, it uses contrarian philosophy. If you or your ad agency are developing an ad and you want it to be truly successful, this is the kind of philosophy you need to adhere to.

6. ***Lasting Impression***

A combination of the above-mentioned factors and very little copy often create a lasting impression.

A number of years ago an ad appeared in Advertising Age that showed a photograph of an egg at the top of the page with a caption that read,

"How do you improve a perfect product?"

At the bottom of the same page, there was a photograph of another egg with a Good Housekeeping seal on it! Period, end of story. No other text, no wall-to-wall text and no superfluous material.

7. ***Graphics plus Photos and Little Text, or the Other Way Around***

Obviously, the example in #6 above clearly indicates that if you tastefully and judiciously select your photographs and graphics and use as little text as possible, you can develop an outstanding ad. This gives credibility to this old saying in advertising: White Space Also Sells.

How to Effectively Position Your Ad in Magazines

Most every advertiser requests the far forward right-hand location. In fact, there has been plenty of research that shows that a great ad will be read no matter where it is placed; whereas, a bad ad will not draw any attention no matter where it is placed. Extensive research shows that if an ad is placed adjacent to related editorial material, this positioning will favorably affect the advertisement's readership.

Size of Ads versus Readership

Extensive studies by numerous publications, including TMC's independent research by Harvey Research and others, clearly indicate that the size of the ad definitely matters when it comes to readership and making an impression on the readers. For example, such research has clearly indicated that a two-page spread advertisement will draw 42 percent more readers than a single full-page advertisement.

Advertising Frequency

Another major area many advertisers seem to ignore (particularly in this economy) is that an advertiser may come up with a small budget and place an ad online or in a magazine a couple of times. Inevitably they then wonder why they didn't get any results.

Assuming they followed all of the above guidelines (but the chances are they didn't follow any of them), an advertisement must appear a minimum of 12 times to:

a) Position the company's product in the mind of the buyer, and
b) Take advantage of the fact that continuous advertising makes the buyer comfortable doing business with the advertiser.

Sustained advertising helps to instill the feeling that the company advertising is substantial enough to still be around in a few years when the buyer may need repairs or service of some kind; whereas, if the ad appears just a couple of times, it has just the opposite effect.

There was a true story in which a manufacturer offered a CTO on a complimentary basis a piece of equipment valued at six figures and the CTO turned it down because he felt that the company might not be around to service it when it was needed. That was simply due to the fact that the manufacturer did not advertise on a regular basis in any trade publications.

Color of the Ads

Selecting the proper color for an advertisement is one of the most crucial elements that will contribute to the success or failure of the advertisement.

Even if you follow all of the above-mentioned guidelines, but you use the wrong colors, the effectiveness of the ad will go down drastically. In a very simplified way, here are a few guidelines for effective color selection:

A. The most effective headline color is red or a warm red.
B. The second most effective headline color is bright orange as long as the quality of color is controlled via a PMS-type ink as opposed to four-color process match.

On the other hand, one must never use red, bright orange, or warm red for small type (eight-point or lower) for text that runs more than two inches wide and several successive lines.

In plain English, avoid using these colors in small print because if you are using them to emphasize something, it will be counterproductive and will have a negative effect because it is extremely difficult to read anything written in red in small print.

In general, black should be used in all small print and the majority of the text because it is the most legible in any size.

The colors to avoid in general are grays, dark grays, dark greens and browns of any kind. As a secondary color, blues, greens or magentas are recommended but rarely as the primary color for text. Yellow is by far the most powerful color when it is used as a screen behind black or red or other colors; that's provided the density or intensity is kept to a very low level.

Indeed, much research has been conducted to show that the use of yellow as a background color increases readability by better than 44 percent. Having said that, one must never try to use yellow as a headline color unless some very dark circumstances are used around it or in the background to make the yellow stand out.

In ordinary printing, yellow must never be used as a headline, text, subhead or anything else. In some cases, yellow could be used provided a black outline is around the yellow to help legibility.

Hopefully, you can see how important the role of color selection is in your advertisement.

A Good Case In Point

A few years ago, TMC's vice president of advertising sales and Nadji attended a convention where an advertiser came up to them and said, "I don't know what we have to do to make our advertisement pay."

Nadji looked at their ad and asked him, "What are you trying to do with this ad?" The advertiser said, "We want to sell predictive dialers, but we are getting no response, and I'd like to know why."

Upon examination, Nadji noticed that the ad did not explain the product the manufacturer was trying to sell. Then he asked the advertiser, "Where does it say what you are trying to sell in this ad?" The advertiser then looked at his own ad and said, "I guess we screwed up."

Nadji told him, "Every ad must have a powerful, benefit-driven headline, but your ad has no headline!"

Then Nadji continued, "There is no discussion of benefit anywhere in this ad that would prompt a reader to call you and ask for information."

Last but not least, the ad had two other problems:

a) It had too much copy that didn't say anything, and
b) It used a repellent color, i.e., dark green, and a pastel shade of green.

In other words, the advertiser had a lousy ad and was wondering why it was not producing any results. As the old saying goes: Garbage In, Garbage Out. Perhaps the biggest crime in the case of lousy ads is that 99.9 percent of the time the advertisers blame the media vehicle although a useless ad will never produce, no matter where you place it.

Four-Color Vs. Black-And-White Performance

According to a Cahners Advertising Research Reports, four-color ads are noted by 45 percent of the readers vs. 33 percent for black-and-white ads. In short, advertising readership increases with the size of ad and the use of color.

The Role of Advertising Agencies

As you consider the complexities of developing a high-quality, effective ad, it becomes clear that often, if not always, the best solution is to outsource to advertising agencies whose core competency is developing advertising campaigns. The agencies know copy, positioning, placement, art and graphics, the needs of the clients and can also find appropriate publications in which advertising could get maximum return on investment.

Having said all of the above, it must be clearly pointed out that an ad agency alone cannot develop high-quality advertising without proper input from the client. Such input must be extremely specific while clarifying whom the target audience is and providing the advantages the products and services offer vis-a-vis the competition.

Remember that if you do not provide proper and complete information to the advertising agencies, then you should not expect a high-quality advertisement.

Hopefully, we have been able to shed some light on the fact that effective advertising takes a great deal of effort, and that a great deal of detail must go into developing an ad that will be remembered and will produce brand recognition and quality sales leads for the advertiser.

Above all, the best advice is to not look for shortcuts because there are no shortcuts in developing quality advertising. If you are not qualified to develop an ad, seek the professional services of a reputable advertising agency.

An example can be cited by describing the success and failure of company X and the ultimate success of company Y.

In the mid to late 80s, company X took advantage of the inbound telemarketing boom by using a toll-free number and advertising it heavily as the preferred source to buy its products around-the-clock. In the early development stages of the company, the firm marketed heavily and practically all day long, every day, until it positioned itself as THE source for the product in question and thus enjoyed the No. 1 position in market share.

A few years later, company X was sold. All advertising, positioning and differentiation was stopped by the new owners.

Company Y came along and did what company X used to do and started to heavily market, advertise, differentiate and position itself as the new leader.

Guess what? Company Y became the unquestionable leader in the marketplace and next to nothing was heard further about company X. This is a true story.

The idea is not to bad-mouth any company, but to simply point out that great marketing, positioning and differentiation made company X successful. But when all of these marketing activities stopped, it lost market share and its leadership position to someone else who did a better job of marketing, advertising, positioning and differentiation.

Why Positioning and Differentiation Are Vital to the Success of any Marketing Campaign for any Product or Service

With so much global competition, customers need a reason to buy from you, and that reason comes from your positioning and differentiation,

which explains to your customer or potential customer what sets you apart or what sets your product or service apart. Without that, no one has any reason to buy your product or service as opposed to your competitors.

***

In their book, "The New Positioning; The Latest on the World's #1 Business Strategy", authors Jack Trout and Steve Rivkin share the following quotes:

A firm in a highly attractive industry may still not earn satisfactory profits if it has chosen a poor competitive positioning.

— Michael E. Porter
The Competitive Advantage of Nations

The key to any marketing plan is positioning.

— Ron Zarrella, Vice President, General Motors
Brandweek

The global marketing imperative: Positioning your company for the new world of business.

Chicago Tribune

More than anything else the success of soft drinks depends on taste and positioning, rather than on presentation and design.

Food Magazine (Holland)

Rolls-Royce to buy Allison, positioning U.K. firm in the U.S.
The Wall Street Journal, Europe Edition

Influencing the Mind

In his book, Mr. Trout defines positioning as follows:

"We have always defined positioning, not as what you do to the product, but what you do to the mind."

Mr. Trout further believes that the ultimate marketing battleground is the mind, and the better you understand how the mind works, the better you will understand how positioning works.

***

Chapter 2

The Rules for Successful Marketing Today

The Dawn of New Media Digital Marketing

Today's marketers face many more options than those of just a decade ago.

Where once there were broadcast, print, direct mail, outdoor and telemarketing, there are now countless digital options. Today's new media marketing vehicles allow your message to be delivered to your prospect in the medium of their choice at the time they are looking to make a buying decision.

Digital or "new media" channels also empower marketers to pinpoint their audiences based on the specific web pages they visit and the specific content they read, resulting in higher ROI and fewer wasted impressions.

Following are the rules you must follow to develop consistent success in your marketing.

Rule #1 – If you're not on page one of major search engines, you don't exist.

Sixty-eight percent of people DO NOT click beyond the first page.

In his marketing classes, Nadji has always said, "if you don't market, you don't exist."

Today, that principle is no longer complete and true simply because of the advent of digital marketing. We all know that nearly without exception, anyone who wants to learn about any topic, the first order of business is to go on one of the leading Internet search engines.

If you are not on the first page of Google, based on your specialty, you do not exist. This is true to even a greater degree for B2B marketers, which are targeting primarily a business audience.

When you do appear on the first page, you will find customers reaching out proactively to you asking about your company's products and services. It works!!

Nadji's company, TMC, has developed the technology and expertise for a company to consistently appear on the FIRST page of Google searches (www.tmcnet.com).

We also know that based on considerable research conducted by various organizations, 68 percent of people do not click beyond the first page. It is further understood that, of the remaining 32 percent, nearly half do not go beyond that and, in fact, look for the first page of listings of a related search term.

The higher the ranking of the executive searching on the web, the less time he or she would have to look for information. Consequently, it is fair to assume that people who serve in a decision making capacity or senior management role hardly have time to go beyond the first page. Based on the above, it should be crystal clear that if you are not on the first page of the leading search engines, in today's new media and digital marketing age, you don't exist.

Rule #2 - Marketing is not a part-time job!

We have seen many entrepreneurial smaller companies wasting their marketing funds by placing an ad or two in a certain magazine and receiving seemingly no business from it. Thereafter, such a company would stop advertising or marketing and claim that advertising or marketing doesn't work.

Worse than that, some ill-advised entrepreneurs and smaller companies have a tendency to cut out all marketing and advertising plans when business slows down. Nothing could be more damaging to an organization than stopping advertising and marketing, which are primarily meant to generate quality sales leads. It is common knowledge that in a slowing economy, many companies could lose as much as 40 to 50 percent of their current customer base. And, if this lost business is not replaced with new business, which usually comes from new qualified sales leads, then it is only a matter of time before such a company would go out of business!

For a complete understanding of the most successful methodologies to generate quality sales leads, refer to *Chapter 3: How to Create Qualified Sales Leads.*

Rule #3 - Warren Buffett's rule: To succeed in business...build a strong relationship with the media.

Warren Buffett, as the world's No. 1 billionaire in terms of net worth and business acumen, practices what he preaches. If you are watching CNBC and Fox Business Channel on a regular basis (as we do and we assume as most business people do) you will have noticed that Warren Buffett appears and is quoted on a regular basis.

As a matter of fact, one might say that the only regular commentator on CNBC, besides the anchormen and anchorwomen, is Warren Buffett. By appearing in the business media with millions of viewers globally, he is not only sharing his experiences with the viewers, but he is also marketing his company, Berkshire Hathaway, in the most effective way at no cost!

To share a similar personal example, the CEO of SIP Print made a habit of regularly visiting TMC HQ offices in Norwalk, CT, to meet with the team and have lunch with senior management. He would even invite his account manager to his home every year. The CEO would also host dinners with key leaders attending TMC sponsored conferences. He was a regular exhibitor and advertiser.

Rule #4 - Do not fight with the media...sure, this is common sense but it's NOT uncommon!

Believe it or not, in spite of the above, we have come across a few ill-advised companies that actually like to pick a fight with the leading media vehicles for practically no reason. Our knowledge of the industry has convinced us that these companies have either already vanished, or it is only a matter of time before they go out of business. One can almost say that if you're not active in the leading media of your industry, you also don't exist.

Rule #5 - To dominate your market, you need to dominate the integrated media.

Going back to the early 1990s, Nadji has been vigorously promoting the concept of integrated marketing. Once again, it seems that he might have been slightly ahead of his time, because the rest of the world has finally realized that integrated marketing, in today's business environment, is the only effective way to reach your market. Consequently, some consulting agencies and business publications have only recently started to admit that integrated marketing is the only way to go. To that we would say, "It's about time."

As students of marketing, we also learned that there is no point in conducting any kind of marketing strategy unless you want to dominate your market. It only makes sense to believe that to dominate your market, you must dominate print, events and online media and there is no shortcut to integrated marketing.

Rule #6 – Long-term marketing success depends upon Customer Care™ and customer interaction.

Communication directly from customers to the company has become a critical component for delivering Next Level Customer Care™, which is the key to customer retention. By understanding that both marketing and customer service are intertwined, it becomes abundantly clear why we need to also focus on marketing.

In our many years of experience in the contact center industry and in our positions as executive advisory board members to the marketing department of a major university, we have come to realize that there is a tremendous need for evolution in the new marketing era, which we call digital marketing. Not only the industry, but also academia, have a long way to go when it comes to the teaching, practice and application of digital marketing.

Put the text in red in a box

All Sales Begin with a Qualified Sales Lead

The job of the sales department is to convert those sales leads into customers. Generating qualified sales leads is the responsibility of the marketing department. If we understand this concept, then we should understand that marketing and customer service, Customer Care™ and customer interaction are all intertwined with one another.

Rule #7 - There is no shortcut in marketing!

A recurring problem we frequently observe, particularly with high tech companies, is that they tend to spend practically every dollar that they have in developing great new technology. However, when it comes to marketing, those same company leaders have said, "We spent $25 million to develop this product, and we basically have no money left for marketing."

Nothing could be more ill-advised. The very principle of marketing says that "if you don't market, you don't exist." Consequently, the people that spent $25 million developing a technology have been wasting their money unless they realize that if they don't market it, the technology would basically become obsolete.

Having said that, it appears many companies are pretty much oblivious to this concept, and they think they can get away with PR only. Once again, that is a huge mistake. These days, there is no way to get a lot of PR in printed publications because many have downsized and editorial pages have become a precious commodity. Consequently, in our opinion, PR alone will never cut it.

Rule #8 - Forget ego marketing, focus on benefit marketing.

One of the leading problems that happens in practically all industries is that the entrepreneurs think they know how to do everything better than everybody else!

They try to act as de facto marketing directors. They place their photographs on their ads (whether they look nice or not) and they prepare an ad around the CEO's photograph where the ad basically says nothing and doesn't give anyone a reason to contact that company for any information about the product they are trying to sell.

In our lifetimes, we have run into many CEOs like that.

One of these CEOs, who fortunately had an incredibly open mind, was receptive to Nadji's advice. He asked Nadji specifically what he needed to do to overcome this problem, and Nadji's answer was, "if I give you a blueprint, are you going to do it?" The CEO said, "If it makes sense, yes."

Then Nadji wrote 12 guidelines for him to follow. He liked the guidelines, and he followed every one of them. To make a very long story short, his company eventually went public, and the CEO who was the founder has now become a billionaire! Every time he sees Nadji, he says, "Nadji, I never forget that you helped me build my company."

Rule #9 - The Right Message to the Right People is the Key to Success.

As a good example of the old adage: Penny Wise and Pound Foolish, certain entrepreneurs who think marketing is a necessary evil give only lip-service to marketing. They might approve an ad that is aimed, for example, at specific manufacturers, but to save money, they place that ad in media that have no direct correlation to their target manufacturer audience.

Therefore, the ad – no matter how effectively it was designed – is not talking to the right people. As a result, the advertisement campaign will fail and, as usual, the advertiser blames the media and not his own poorly conceived actions.

Rule #10 - Don't try to build a better mousetrap...build what the market needs.

Believe it or not, this is also another very common mistake. Creative entrepreneurs often build a better mousetrap only to find out the mouse died 15 years ago.

The point is that the product should exist because the marketplace demands it and not the other way around! These short-sighted marketers do things backwards by preparing a product that the market doesn't care about, and they waste a ton of marketing dollars to promote it. After all the money is wasted, they realize that they are doing everything backwards.

Rule #11 - Focus on quality and build what customers need, not what you think they need!

When it comes to quality, remember Henry Ford's great comment,

"A good design sells the cars, but good quality brings the customers back."

We wonder how many manufacturers have actually heard of this. One of the reasons that the Japanese and the Koreans and the European car manufacturers gained a tremendous market share from U.S. manufacturers is the U.S. automakers neglected quality. As a result, some of the U.S. car manufacturers went through bankruptcy.

Every foreign car owner that you ask the question, "Why did you buy a Japanese, Korean or European car?" will answer "better quality." What we don't understand is why it took Detroit so long to figure out, through simple market research, why it lost so much market share in the meantime?

Today, we recognize improvement in the quality of U.S. automakers, and a number of recalls have been announced by foreign manufacturers, so hopefully we are seeing a rebound for the U.S. brands such as General Motors. Steve drives a GM made car and is extremely happy with its quality.

Rule #12 – Focus on Awareness

As indicated above, one of the greatest weaknesses of technology companies is the notion that "our product is good enough and people will find out about it," by osmosis we suppose!

It should be crystal clear that no one will buy anything from you unless all concerns are answered, and the only way to do that is via integrated marketing.

Rule #13 - Positioning and Differentiation

We need to understand that today in this super competitive marketplace, if you do not give a reason to your potential buyers why they should buy your products, they simply will not choose your products over your competitor's product. In other words, you must find what makes your product unique and translate that in terms of benefits to your potential buyer and then communicate the benefits and the differentiation factor to the marketplace 24/7. In other words, positioning and differentiation are not part-time jobs. You need to do it every day, every minute, every month, every year to be successful.

Rule #14 – Focus on Relationship Marketing

Even if you apply all of the 36 marketing principles detailed at the end of this chapter, if you do not have a well-natured relationship with your customers and your customer base, you may face great difficulty selling your products and services. The reason is that 75 percent of buying decisions are made based on emotion. In other words, even if you have the best product in the marketplace and your competitor has a better relationship with your prospect, 75 percent of the time, your prospect will buy your competitor's product.

Therefore, you need to do everything humanly possible and affordable to build a solid relationship and a continuous relationship with your prospects and customers. Otherwise, all bets are off.

Rule #15 – Avoid AF, Build a Functional Advertising and Marketing Campaign

I suppose you are wondering what does AF stand for. AF basically stands for artsy fartsy in advertising.

One of the greatest mistakes made in advertising is that complete control is given to the creative people to design an advertisement that looks extremely pretty and perhaps beautiful, but it doesn't say anything and it is not functionally effective. Seasoned marketing people will tell you that there is much more to developing an effective advertisement than making it look artsy fartsy.

AF ads may win advertising awards just based on looks, but when it comes to productivity, they are practically useless unless good copy and graphic communication skills are blended with uniqueness and convincing creativity in the copy. If you would like to take your advertisement to the next level, you need to be so creative that it would become a memorable advertisement for years to come.

Rule #16 – Try to Outsmart, Outthink and Outsell your Competition

Obviously, this is much easier said than done. However, if you truly are interested in dominating your marketplace, there is no shortcut to this rule.

Not only do you need to outsmart, outthink and outsell your competition, but also, you need to dominate online, print and social marketing. This is the most effective solution to market domination that exists.

With so many global competitors, being a copycat will not get you anywhere. So you need to outsmart your competition and be original.

To truly outthink and outmarket your competition, you must dominate the three areas mentioned above such as online, in print and in person, such as at trade shows.

Then, if you do all of that, you should be on your way to success, provided that you have an effective sales department that can effectively bring in the orders.

In other words, even if you do everything requested above and your sales department is extremely weak, nothing will sell and all marketing dollars are wasted. So there is a cause and effect relationship here. Without effective marketing, there will be no qualified sales lead generation, and without qualified sales lead generation, the sales people will have nothing to sell with.

Rule #17 – Be a Peacock in the Land of Penguins. Don't be a Copycat.

Every company wants to be a peacock in the land of penguins. So appreciate the importance of being different than your competition.The copycats, at best, may only get the crumbs; and the crumbs are not going to sustain a viable business. Having said that, there is no other way to gain market dominance but by having an outstanding product and marketing it like a peacock in the land of penguins.

Of course, it is easy to be a penguin; they all look alike and they all are a copy of one another. Obviously, it is extremely difficult to be a peacock in the land of penguins, but it is not impossible and when you achieve it, we are sure you will enjoy maximum market share and you will find that all of the difficulties were well worth the effort.

Rule #18 – Think Outside of the Box

Along the lines of being a peacock, one must think outside of the box and avoid following the trend or should we say the competitive trend. Thinking outside of the box requires extreme creativity and vision. Obviously, not everybody has that, but even if you do not currently have that type of talent in your marketing department, you need to hire people that are blessed with such attributes. Thinking outside of the box will be difficult, indeed, but it will set you apart from competition and that is vitally important today if your marketing department is going to be marketing your products and services.

Rule #19 – Remember the Golden Triangle Rule

This rules says, you need to dominate print, online, and in person at events to dominate your market.

Many marketing departments are extremely restricted by budgetary problems. Limited in this area, they choose only one of the above vehicles. In other words, they only choose print or online or events and that will never work. Nadji is familiar with a company that existed in the Chicago area a number of years ago and did not believe in anything but print advertising. Today, that company barely exists.

At the same time, there was another company also located in Chicago not far away from the first company, which believed in dominating the market place and developed an integrated marketing strategy accordingly. That company became extremely successful.

In other words, don't look for shortcuts. Dominate all three areas of marketing with an effective, outside-the-box-thinking marketing plan and you will find that no competitor can take anything away from you.

Rule #20 – Innovate, Don't be a Copycat

Please forgive us for being slightly redundant in some of these. One of the greatest rules of direct marketing is that if something is important, repeat it several times in your marketing and advertising copy because executives are extremely busy and they only skim a marketing piece or advertising piece and if you don't repeat it a few times in different ways, they will probably miss the point. Therefore, it is in this spirit that we keep driving the importance of innovation and avoiding being a copycat.

Along those lines, Nadji's father had a good point regarding the copycat. He said that, "If the Good Lord wanted us to be copycats, we all would have been created as monkeys!"

Rule #21 – Focus on Strategy: Seat of the Pants Marketing Doesn't Cut it Anymore

We love entrepreneurs. They are really and truly a different breed of cat. They all have tremendous egos.

If the ego is out of control, they get nowhere. But if the ego is managed and properly directed, they will go everywhere.

In the call center industry, we remember two executives, both extremely intelligent and blessed with powerful egos. In the first case, the ego was under control and that person is a billionaire today. In the second case, the ego was not under control and even though that person had greater business skills, he only became a multi-millionaire. So the ego can be a double-edged sword. If you don't have it, you cannot have outstanding achievements, and if you have it and it's out of control, it will definitely hurt you in the long run.

Having said all of that, one of the problems with most entrepreneurs is that they think they know everything better than most, and therein lies a major problem for their growth. Indeed, many of them depend on seat-of-the-pants marketing, which may have worked years ago, but it doesn't work today. Marketing must be sophisticated, be well planned and well strategized to be effective; once again, there is no shortcut in these guidelines.

Rule #22 – Focus on Quality Lead Generation and Follow Up

Lead generation is the forgotten art in marketing. Lead generation indeed plays a vital role in business. Suddenly, companies that didn't seem to care about lead generation as recently as a few years ago are all interested in lead generation today. What makes matter worse is that the universities also do not teach or explain much of anything regarding lead generation to the students.

That is a crime because just as "companies live or die from repeat business", by the same token you can also say that companies live or die from quality lead generation or lack of it.

Suddenly, corporate America has woken up to this very important fact.

The other problem in lead generation was that in the past 10 to 15 years, many companies would spend a lot of money attending trade shows, marketing and advertising to generate leads, and 80 percent of the time the leads were not followed up on or they were given to a very poor sales staff which could not generate any business from them. We thought that was a disgrace. But today, we are happy to say that many companies have come a long way and are working toward quality lead generation on a regular basis.

Develop a Full-Time, Integrated Marketing Plan

Many companies could be doing a better job of basic marketing. Although learning the basics of marketing is critical, to be truly successful, companies need to become proficient at full-time, integrated marketing. Unfortunately, the practice of integrated marketing is not widespread.

What does a full-time, integrated marketing program entail? It would incorporate a regular, systematic, full-time approach to marketing in all of the following twelve areas, including:

- Effective digital marketing
- Consistent public relations efforts
- Effective social marketing
- Regular editorial publication of success stories
- Effective and regular advertising with supplements and inserts, etc., for maximum results
- Effective trade show exhibiting (including effective pre-show, during-show, and post-show marketing)
- Effective implementation of technology

- Effective relationship marketing with a focus on customer experience
- Effective Customer Care™ and customer service
- Effective data analytics, including modeling, profiling and customer journey mapping
- Effective use of online media
- Last but not least, effective use of call center resources

Integrated marketing is indeed the heart and soul of effective, full-time marketing in any corporation – if the corporation is to survive global competition, and meet the competition head-on in doing an effective job of market-share protection and market-share augmentation.

Part-Time Marketing Is Widespread

Unfortunately, part-time marketing is a disease that has infected many corporations, large and small. Those that have been able to cure this disease and put in place a powerful, high-impact, integrated, full-time marketing program will usually prosper, while those that do not, even with a far superior product, will not go anywhere. You can look around you in any industry and notice that this is a true statement time and time again.

Part-time marketing is particularly prevalent in companies where the chief executive and all other members of the organization wear many hats. In these types of environments, it is very easy to get bogged down in day-to-day operations and completely neglect marketing opportunities for your company. If you look around you will find that this is the rule rather than the exception among entrepreneurial companies. Such companies not only get entangled with solving day-to-day problems, but they also somehow prefer to admire and enjoy what they have developed and think the rest of the world will buy their product simply because they happen to like it.

In reality, the market is not made for your products, rather, your products are made for the market. If the market does not know about your product, it has no reason to purchase it.

Companies such as those just described every now and again, due to pressure from their sales departments, may take a feeble action toward marketing. Entrepreneurs often think that if they erratically show up at conventions or erratically place advertisements or erratically participate in editorial opportunities, they are doing a good job of marketing. In reality, nothing could be further from the truth.

A marketing organization that prides itself in doing a first-class job must keep its name constantly, even daily, in front of customers and in front of the industry. Considering that exposure brings familiarity and familiarity brings comfort in purchasing, if a company is erratically marketing or if a company is out of sight due to part-time marketing, then the buyers' perception is that the company is not, in fact, fully committed to the industry.

Periodic support of the industry or periodic exposure to the industry indicates periodic commitment to your products and services, and to the industry. Therefore, no one would consider your company as a serious vendor!

Buyers also perceive erratic, part-time marketing efforts as a sign that a company's products and/or services must be inferior to its competition, whose message is constantly in front of them. Whether or not the buyers' perception is true makes no difference because, in marketing, perception is everything! And perception in the mind of the consumer is reality, like it or not.

The way in which a company positions itself will define the perception consumers have of that company. To create a positive impression, a company must commit to an all-encompassing marketing plan that is executed on a consistent, steady schedule. To ignore this principle is to allow a vacuum to be created for your competition to say, "Yes, Company Z has inferior products and, in fact, may not be promoting itself due to financial problems."

In other words, if you do not take the time to position yourself and carve out the positive perception you desire, your competition is given the opportunity to carve out a position and perception for you!

There are 12 crucial ways to combat negative word of mouth and consumer perception, as we identified above. Let's expand on three of these areas: public relations, editorial exposure and effective advertising.

Public Relations

The first, and most basic, step in your marketing plan is to establish strong public relations. The best way to do this is through the preparation and regular mailing of press releases to the appropriate media describing your products, services and success stories.

Well-written and targeted press releases are one of the most effective and free methods of awareness creation about what your company does. In addition, press coverage of your company's milestones can serve as a great motivator to your company's employees. Unfortunately, we have learned that the power of the press release is often overlooked.

Nadji addressed a convention consisting of an elite group of top management from small to mid-sized companies. During his speech, he asked the audience, "How many of you regularly provide the press with literature and updates about your products, services and corporate achievements?" To his surprise, only one out of a group of about 100 executives raised his hand!

The question is, "Why?"

A well-written, well-targeted press release is one of the most effective – and cost-effective – ways to promote your company and establish the position and perception you desire.

Editorial Coverage

Not only do too few companies issue appropriate press releases on a regular basis, even fewer avail themselves of editorial opportunities available through the major media outlets and publications in their industries. Of course, respected media sources do not allow their editorial content to serve as commercials for vendors, but by preparing authoritative, how-to pieces on the subjects in which your company bases its research and development, you will convey your authority and expertise in the field to the reading audience.

To be successful at editorial placement, be sure to follow these guidelines:

1) Query the publication's editors in advance about the relevance of the topic about which you plan to write.
2) Obtain and strictly adhere to the publication's writer's guidelines.
3) Meet the requested deadline.

In addition to preparing articles for publications, make a concerted effort to stay abreast of the magazine's special roundups, awards, buyer's guides, supplements, etc., and be sure to participate.

There is one caveat, however, to editorial participation: Be sure you are dealing with reputable publications before you spend any time or money preparing editorial materials for them. You clearly want to deal only with publications that make a strong distinction between editorial and advertising, as they are the only ones that readers respect. In other words, beware of the publication that tells you it will run your article if you place an ad, or the one that tells you it will run your article just because you are a frequent advertiser.

Advertising

Effective advertising is more than just listing your products and services. You have to give your reader a reason to buy from you as opposed to buying from your competitor: if your ad does not do that, you should discard the original ad and prepare another one. You may find it necessary to switch to a more effective advertising agency.

Not only must your ad be effective, but as we've said, it must appear frequently in front of your targeted audience. We once heard a businessman say, "We placed two or three ads in X magazine last year and we didn't see very good results, so we won't advertise anymore."

Obviously, if the targeted audience was there, either the frequency of the ad placement was not enough or the ad copy and layout were ineffective, or both. If the advertisement is not effective, and if the frequency is not there, what reason does anyone have to buy from you? Here again, the factors of frequency and positive perception affect consumers' image of your company. If you advertise in a poor manner, you are creating a perception that your company is less than it really is.

The Bottom Line

Once a product or service is positioned effectively, every piece of the integrated marketing program must repeat the company's position at every junction: public relations, direct marketing, call center, trade show marketing, digital marketing, social marketing, relationship marketing, etc.

If you don't speak about your products and services, and how they differ from your competition, others will speak for you, and often in an unfair manner.

Now is the perfect time to further develop your marketing strategy that is full-time, effective and integrated with all the essential components. Make sure everyone in your company understands that marketing is not a part-time job!

BONUS MATERIAL: THE 36 PRINCIPLES OF SUCCESSFUL MARKETING

1. If you're not on page 1 of major search engines, you don't exist. 68 percent of people DO NOT click beyond the first page.
2. Marketing is not a part-time job!
3. Warren Buffett's rule: To succeed in business...build a strong relationship with media.
4. Do not fight with media...sure this is common sense but it's NOT uncommon!
5. To dominate your market, you need to dominate the integrated media.
6. There is NO shortcut in marketing.
7. Forget ego marketing...focus on benefit marketing.
8. Right message to the right people is THE key to success.
9. Don't try to build a better mouse trap...Build what the market needs.
10. Focus on quality and build what customers need, not what you think they need.
11. Focus on awareness.
12. Focus on positioning and differentiation. You cannot succeed without them.
13. Focus on relationship marketing – because 75 percent of buying decisions are based on emotion.
14. Avoid AF (artsy fartsy) – build a functional advertisement.
15. Try to outsmart, outthink, out market and out sell the competition. It's easier said than done!
16. Be a peacock in the land of penguins. Don't be a copycat.
17. Think outside of the box.
18. Remember the Golden Triangle: Dominate print, online (social media) and events to effectively lead your market.
19. Innovate, don't be a copycat!
20. Focus on strategy; seat of the pants marketing doesn't cut it anymore.
21. Focus on quality lead generation and follow up.
22. Create awareness about your company via integrated marketing.

23. Focus on powerful branding.
24. Respond to the objections of the buyer (man on the chair ad).
25. Focus on every detail.
26. Choose the right colors and avoid the wrong ones in your marketing.
27. Use spectacular graphics to ensure your message is inviting to read.
28. Use powerful, benefit-driven headlines and copy with the right colors.
29. In marketing, timing is everything.
30. Focus on quality lead generation.
31. Create an outstanding perception and back it up with quality performance.
32. In marketing, perception is reality.
33. Implement the highest quality, superb customer service and Customer Care™.
34. Above all, under promise and over deliver.
35. Focus on award marketing. Nothing is more powerful than a third-party endorsement or validation.
36. Focus on electronic/digital marketing.

Chapter 3

How to Create Qualified Sales Leads

LEAD GENERATION: The Forgotten Link in Savvy Marketing

Why Lead Generation?

If you follow the anatomy of a healthy organization, you will find that without exception, no company can exist without new business and, simply stated, no company can remain in business without sales. It follows, therefore, that to generate sales, one must have sales leads because all sales begin with sales leads.

As vital as lead generation is, it is mind-boggling that so many companies ignore this phenomenally important part of business and simply give it casual attention, if any at all.

Sources of Leads

Leads can be generated from any or all of the following:

1. Trade shows
2. Print advertising
3. Call Centers
4. Social marketing
5. Online and digital advertising
6. Direct mail
7. Integrated marketing (which is regarded as the most powerful method)

8. Effective response-driven campaigns (which begin with response-driven advertising)
9. Effective positioning (no marketing campaign could be functional without it)
10. Differentiation (again, no marketing campaign could be functional without it)
11. Public relations

Without positioning and differentiation, the customer has no reason to buy from you, and in this extremely competitive environment, there is no room for mistakes in this area.

Having said all of the above, keep in mind the following foundational principles for marketing success.

Marketing Is a Full-Time Job

We said it before and we'll say it again: There are many misguided, so-called marketing managers or CEOs of mom-and-pop companies who think of marketing as a necessary evil. Such companies are doomed to fail because you simply cannot exist without marketing. In plain English, marketing is not a part-time job. If you can't market full-time, just don't do it at all and simply get out of business.

Creative Marketing

With mass-marketing's proliferation of advertising messages, the ones that will get consumers' attention are those that are extremely creative and attract the recipients' attention immediately. Such messages and marketing programs are usually benefit-driven so that consumers have a reason to reach out proactively to the company and request the product or service.

Innovation and Marketing

Blend innovative marketing and creative marketing together because they both seem to be unusually effective in attention grabbing.

It is increasingly normal for the techies and the engineers to think of innovation and new technology and totally ignore marketing. At best, they usually consider marketing a necessary evil. Obviously, companies that put people with that mindset in charge of business decisions will not go very far.

Lead Sourcing Is Vitally Important

The determination of the source of the lead is vital to your marketing success.

In any marketing program, you MUST be able to determine the ROI. Obviously, you cannot determine the success of any marketing campaign if you are not able to determine the source of leads. Once again, you must follow up on all qualified leads immediately, otherwise they are useless.

Timeliness

Sales leads need to be followed up in a timely manner, as well. Two-week-old leads are usually useless.

The Role Of Customer Care™

Next comes the job of Customer Care™, the objective of which is to keep the customer satisfied by developing a strong relationship with the customer. Through this relationship and making your company indispensable to your customer, you can count on keeping that customer.

In short, the job of advertising is to generate sales leads, and the job of salespeople is to close the sales and turn the leads into customers. The job of CRM and Customer Care™ is to keep the customers.

The Most Effective Way to Generate Leads

Integrated marketing and/or multimedia programs are effective ways to market and generate sales leads. Customers tend to react to advertising and marketing materials in different ways. In other words, some prefer voice (radio or telephone), others prefer video (online or television), others magazine and print advertising, and still others social and digital channels.

To conduct a winning marketing program, you must consider integrated marketing as the vital point of your marketing program, because the one-size-fits-all approach does not work in marketing.

The Nature of Incoming Leads

There has been an evolution in the nature of incoming leads. As opposed to leads from direct mail via coupons, postcards, regular mail and response cards, the nature of incoming leads generated from a combined online and telephone response strategy are more than 80 percent more effective.

No matter where you advertise, today more than 90 percent of the leads are coming via the phone and digital channels.

The Quality of Salespeople and Timeliness

You also need an automated process to make sure the leads are immediately tracked in a database. They must then be e-mailed and/or called and/or mailed information, then followed up as appropriate by a salesperson.

If you don't keep in mind all of the above guidelines, such as appropriate integrated marketing, etc., you still may hit a point where your marketing campaign is not producing the desired results. Develop a checklist to determine where there is a shortfall and disconnect in your marketing campaign and fix it.

Remember that every sale begins with a qualified sales lead. The responsibility of the sales department is to convert those qualified sales leads into customers.

If your marketing team generates unqualified sales leads, there will be tremendous and expensive waste throughout your organization. The resources allocated for purchasing lists, technology, payroll and other related details might as well be dumped in the trash.

There is a plethora of channels companies can use today to generate qualified sales leads. For B2B experiences, the most valuable and effective tools for gaining business organization customers, are:

1. Exhibiting at targeted industry events
2. Webinars – In our judgement webinars are the most effective way of generating B2B sales leads today – provided they are handled under the proper guidelines.

Let's say a company creates a new software to increase business productivity by 25 percent.

The headline for the webinar announcement must say something like:

25 Percent SAVINGS IN BUSINESS IS A REALITY
With Product XYZ

Then the company would effectively conduct a demonstration during the webinar of how the product works and how it increases productivity resulting in the 25 percent savings.

The reason for conducting the webinar is to generate qualified sales leads. The people that sign up and attend must already be interested in the product or they would not waste their time. They will be enthusiastic in the topic and therefore will be qualified sales leads.

<u>How do you get the prospects to become aware of the webinar?</u>

The key to identifying an audience is to find a partner who has an existing audience of potential sales leads. Once you find the leading partner, why would you go anywhere else if they have a pipeline to your potential customers?

A Fortune 100 company developed a benefit-driven webinar for TMC's audience of C-level executives targeting a cross-section of industry organizations. The initial live webinar event generated 1,500 qualified sales leads. Then, after placing the content of the webinar in the TMC archives, the company generated an additional 1,500 qualified sales leads within a few months after the initial airing of the webinar, resulting in a total of 3,000 qualified sales leads for the company!

Content Marketing

One of the relatively recent concepts in marketing is content marketing. Content marketing is the marketing and business process for creating and distributing relevant and valuable content to attract, acquire, and engage a clearly defined and understood target audience – with the objective of driving profitable customer action.

Readers should be referred to a phone number or web address to generate qualified sales leads. There, subscribers are introduced to your company. Again, you need to do your homework to define benefit-driven headlines. Content marketing will continue to be a major part of marketing going forward. Remember, without qualified sales leads, no company can exist.

Advertising for sales leads is often one of the biggest spending wastes of any company. If your marketing manager does not understand the importance of developing and following up on qualified sales leads, you are wasting precious company resources.

(AND, your marketing manager could be a WI for your business. Refer to Chapter 9: Hire for Marketing Aptitude, Not Just Experience. The 44 Characteristics of the Right People, for the definition of WI.)

To find out if your marketing leaders are competent and have potential, require them to pass our proprietary Test for Marketing Ability™ provided in Chapter 10.)

Whitepapers

A whitepaper is an informational document offered by a company to promote or highlight the features of a solution, product or service. Whitepapers are sales and marketing documents that help to persuade potential customers to learn more about or purchase a particular product, service, technology or methodology. Whitepapers are an extension of your content marketing strategy and are helpful in creating qualified sales leads.

12 Guidelines for Developing Effective Marketing Campaigns

1. Make sure that your headlines and titles for your advertisement, webinar, whitepaper, online or social promotion, or any content marketing project, are as enticing as possible.
2. If lead generation is important to you, then your whitepapers, case studies, webinars, etc., must be informative and tactfully response driven with a strong call to action.
3. **Your titles, keywords, etc., must be short, to the point, timely and benefit-driven.** You must cut through the clutter to gain attention.
4. If your headline and content are not enticing and do not meet the above criteria, you will NOT get the desired result. In fact, you may get no results at all.
5. Remember that any decision maker's time is precious. They don't have much free time, and they will not take the time to read boring copy, vague titles, or vague anything else and try to figure out why they should contact you for further information.
6. One good way to test a good headline for your online or other marketing strategy is to ask yourself the following question:

   "Is there anything in this headline or this title that would inspire me (the prospective buyer) to spend my precious time reading what I have just produced?"

   If not, change it, rewrite it and make it benefit-driven. Otherwise, no one will read it, and you will end up blaming whatever media you are using, which is almost like barking up the wrong tree. We are sorry to be so blunt, but after many years of marketing leadership and teaching marketing, we simply cannot afford to mislead anyone and therefore, we must tell it like it is.

7. If you are not willing or able to take the time and do it right, it is far better not to do it at all and waste a lot of money and blame the media for poor performance.
8. **The key to promotion of your product and services is via MARKETING THROUGH EDUCATION. There is absolutely no shortcut for this.**

   Informational presentations by nature can be boring. They have to be written in an entertaining and engaging manner, otherwise you will lose your audience before you have created any interest. There is no substitute for **MARKETING THROUGH EDUCATION** when it comes to selling, promoting or marketing, especially for technology-related products and services.

9. **In Marketing Timing Is Everything**. This is particularly important in the marketing of events and webinars. The longer the time you commit to market and promote your event, the greater your audience. A minimum of two months is required to market an event effectively.
10. The contents of whitepapers and webinars must contain truly new ideas, information, unique applications and, above all, they must offer a viable solution to your customers' or prospects' REAL NEEDS.
11. No one wants to read a whitepaper or attend a webinar to learn about a "so what" topic or content.
12. The above guidelines are based on years of marketing experience and they are meant exclusively to help you succeed. The rest is up to you.

If you don't follow the guidelines you will invest valuable resources and not receive a return on investment. You must prepare and plan. Be ready to go. Like anything else, you get out what you put in. Benefits must clearly be indicated.

Producing Qualified Sales Leads Is the Lifeblood of Every Organization

Online Marketing

Otherwise known as digital marketing, online marketing is obtaining a tremendous increase in market share.

The keys to online marketing are as follows:

1. –Webinars, as we've discussed, to generate qualified sales leads
2. –Online Communities - extended coverage online of your products and services
3. –Social Marketing - Facebook, Twitter, LinkedIn, Instagram, etc.

Integrated Marketing

As the name implies, integrated marketing involves the combination of the specific marketing channels in a coordinated campaign.

1. **–Use Benefit-Driven Headlines – REPEAT; otherwise, no one will read the text.**
2. **–The most important concept for the first line of the copy is as follows. The first line of the copy is more important than the next one thousand words. If the first line doesn't tell your audience WHY to read it, they WILL NOT read it.**

Marketing Blunders: When Will Business Leaders Ever Learn?

It never ceases to amaze us, but we discover marketing blunders almost on a daily basis all over the place! The problem is not only the fact that these blunders could be corrected, but often it appears no one even makes an attempt to try to fix them.

We pointed out that many marketing managers could not even pass an elementary marketing test involving defining marketing adequately.

Having said that, we would now like to focus on many of the marketing blunders that are occurring on a perennial basis.

We have decided to provide a summary of some of the most common mistakes made in marketing on a case-by-case basis. As you will notice, many of these cases are probably the same experiences you have also had in the past.

The Biggest Mistake of Them All: Stopping all sales, marketing and advertising plans in a slowing economy

Savvy marketers know that the best time to market and expand their market share is when the economy is at or near a recession because that is the time when most other marketers (or the CEO or CFO behind those marketers) decide to cut all marketing, advertising and sales-related travel budgets.

This kind of inexcusable and downright stupid action is like a doctor trying to save a patient's life, but first cutting off oxygen to the patient's lungs by sealing up his/her nostrils and taping up the mouth and wondering why the patient died! These idiots (and we are sorry, but we think they deserve to be called that) don't realize that in a down economy, a typical company loses between 50 to 75 percent of its customers due to natural attrition and due to economic forces beyond their control.

The only way to remain in business, in that case, is by bringing in new business. That's right, you must bring in new customers. And new customers come from new sales leads. And new sales leads come from advertising, marketing, promotions and trade show exhibitions. This is obviously very simple, clear logic. It is as logical as two times two equals four. Yet, it is amazing that 80 to 90 percent of marketers either don't seem to realize this fact, or their hands are tied by the CFO, CEO, COO or someone else in the organization who doesn't understand marketing.

It's mind boggling, but that's how it is. And that explains why budgets for advertising and marketing promotions have been slashed to pieces, even within companies that should know better, including some bellwether, high-tech companies. We sometimes feel as if we are speaking to a wall, but we do hope these people finally get the point and start promoting their companies, products and services, because if they don't market, they don't exist.

60 To 70 Percent Of Sales Leads Are Never Acted On!

This is the second biggest problem: Some companies spend millions of dollars advertising, exhibiting and sponsoring conventions to generate new sales leads, yet 60 to 70 percent of the leads are never acted on! There ought to be a law against this. The question is, if you are not going to do anything with the sales leads, why bother advertising or exhibiting at the conventions, then turn around and blame the digital portals and magazines in which you advertise or the conventions at which you exhibited for not bringing the right people to the show or for not generating quality leads?

A Sad Case In Point (This is a true story)

A leading company, Company Y, advertised in Magazine X for a decade and always expressed major satisfaction with all the results. In the eleventh year, the company suddenly stopped advertising. The sales manager from Magazine X called the client and asked why they stopped advertising. The client replied, "Because we are no longer getting sales leads from our advertisements!"

About a year later, the sales manager of Magazine X ran into the client from Company Y. They met over lunch and the topic of Company Y's advertising came up. The client from Company Y confided, "I want you to know that I just fired my secretary!" The sales manager asked, "Why?" The client replied, "Your magazine had been producing sales leads all along during the last year we advertised; however, my secretary was putting all the sales leads in a file and accumulating them for 12 months without telling anyone!

"We simply discovered this unfortunate situation by accident, and you should be happy to know that she no longer works for the company and we are pleased to start advertising in the next month!" Need I say more?

Lack of Persistence

In this age of e-mail, voice mail and highly automated communications, it has become ever more difficult to directly reach decision makers. Consequently, about 10 percent of leads are wasted because salespeople simply make a single communication attempt and if they don't receive a response from the decision maker, they don't try again!

The other problem is that many people let the leads pile up (as the secretary mentioned above did) and they become totally worthless, yet they are given to salespeople to make cold calls! This not only discourages salespeople, but also helps them get in the habit of ignoring sales leads. A third category of problems along these lines is when companies do not do a diligent job of assessing the quality of sales leads and, therefore, create further frustration for salespeople, which wastes time and money!

No Lead Tracking!

Another reason many marketers fail is they simply do not do any lead tracking, thus they have NO idea what marketing plan works or doesn't or what advertising campaign works or doesn't. This is sure to produce millions of dollars of waste in marketing budgets.

Meaningless Leadership and Innovation

There are many companies whose executives believe that if they are visionary and highly innovative as a company they don't need to advertise or market their product. Nothing could be further from the truth. Innovations mean absolutely nothing if you don't market and position yourself as a leader and an innovating company.

The Entrepreneur's Greatest Mistake – Not Believing in Advertising

The history of corporate enterprise is loaded with companies that have lost the leadership position because the executive(s) at the top does not believe in advertising.

A classic case in point is Henry Ford's blunder (with all due respect to his genius as the first automobile maker to recognize the benefits of mass production). It has been said that Henry Ford was completely against advertising, marketing and promotions because he thought if you are first in your business, then there is no need to advertise.

He could not have been further from the truth, because he eventually lost his leadership by being totally oblivious to the needs of people (as evidenced by his famous and unfortunate comment that, "People can have my Ford cars in any color they like as long as it is black!"), and the need for marketing, advertising and promotions.

In other words, he told his customers, "I am doing you a favor by selling you a car and because I am the only one in the market, you have to take it, as I like it!" Then along came General Motors, taking major advantage of Ford's blunder. GM cars came in multiple colors, with different options, and it backed its cars with powerful marketing and advertising.

The results are very clear; General Motors has been the No. 1 U.S. automaker and Ford has been the perennial No. 2. Always a bridesmaid, but never a bride!

If It's Free, I Don't Want It

You would be amazed at how many companies ignore the highly effective opportunity of being in printed or online buyer's guides, which are normally vehicles of free marketing support. It is amazing that some companies don't even bother to send free listing forms to be included in a buyer's guide, but then pay for an advertisement in the same issue! We are sure you will agree with us that this simply does not make sense, but believe us, that is how it is.

TECH IPO BLUNDERS

There have been numerous examples in the last 20 years of technology (and other) companies that have gone public with promises of gold at the end of the IPO rainbow. Some were legendary successes, but many of the organizations soon collapsed. While analyzing the reasons behind it, we have come to three conclusions:

1. The extreme pressure by Wall Street for rapid growth in sales and profits has contributed to the problem.
2. The fact that the owners and founders of the companies were able to obtain considerable amounts of money by selling their shares of stock left little incentive for them to stick around and run the companies the way they had run them prior to going public. Therefore, a group of inexperienced, recent-grad MBAs (who didn't have the experience to run a convenience store at a profit) came along and started running these companies. Unfortunately, they didn't understand the complexities of their industry, and as industry veterans attest, unless you know it inside and out, you don't have a chance of being remotely successful running it. Nevertheless, many more jumped on the bandwagon, went public and the rest is history.
3. All of the troubled companies had one thing in common: NONE OF THEM HAD EVEN A MEDIOCRE MARKETING AND ADVERTISING PLAN!

Many investment bankers, attracted by the growth of an industry, would come in with zero knowledge and start buying incompatible companies, forcing them to work together. As we all know, oil and water do not mix. The result was a disaster.

Perhaps the stupidest of all these deals in the call center industry in particular, was that a group of investors acquired two companies, each with a great reputation, but with incompatible corporate cultures, and those investors tried to converge them. Of course, it didn't work out because of the difference in philosophies.

Believe it or not, they completely eliminated the name of these companies and created a new name that had absolutely no meaning, no value and was impossible to remember. By now it must be mindboggling to you how stupid this whole transaction was.

But wait, it gets even worse. After they changed the name of the company to a stupid name, they stopped all advertising and did not tell anyone that this ridiculous name represented the combination of two previously respected and most prestigious companies. They thought that the whole world would automatically know that this acquisition had been made. They lost millions of dollars, and the company was soon on the verge of bankruptcy.

If *Harvard Business Review* is looking for the stupidest mistake made in corporate America, this would have to be right at the top of the list. In short, the investors took two great companies and ran them into the ground without having a clue because they had no marketing and advertising concerning what they were about or why they existed, and no one would do business with them!

IN SUMMARY

When you look at the numerous mistakes made in corporate America when it comes to sales, marketing and advertising, you have to wonder, **"When will they ever learn?"**

If the objective is to waste a ton of money and generate a lot of sales leads and then do nothing with them, or place poorly prepared advertisements that are counterproductive or acquire companies without knowing what you are doing or lose market position and leadership position due to ill-advised opinions against advertising and marketing, one has to wonder how long corporate America is going to put up with this sheer stupidity and wasteful or no spending in marketing departments.

If the objective is spending a lot of money and getting nothing for it, many are doing a good job. But if you are looking for results, those types we questioned should be replaced with people who really know what they are doing and those who really care about their company's prosperity.

Chapter 4

Award Marketing and Strategic Competitive Intelligence

AWARD MARKETING: The Most Effective Way to Differentiate Your Firm!

Award Marketing Goes Beyond Hanging It On The Wall!

Awards without Marketing Are Meaningless

In today's super competitive environment survival of the fittest often boils down to third-party endorsements!

Competent marketers know that beyond such crucial criteria as quality, value, CRM, price, service, etc., which practically every company claims, it is vitally important to rely on third-party endorsements to differentiate themselves from the crowd!

A coveted award earned from a reputable source can be the most powerful tool for differentiation by far, provided that:

a) It is bestowed from a reputable industry-leading source, and
b) It is marketed effectively.

One surefire way to differentiate yourself from competitors is to prominently display and promote the award logo. Awards are normally presented to companies that have reached the upper echelon of their respective industries. What better way to tell potential customers that your organization, your products and your services are a step above the rest?

To help you maximize the impact of an award logo, we've compiled a quick list of potential uses. Be sure to co-locate your company logo or product logo with the award logo and use them on:

- Your Company's Website
- Trade Show Displays
- All Advertisements
- Brochures
- Business Cards
- Company Letterhead & Envelopes
- Incentives & Gifts for Customers and Employees, such as pens/pencils, clocks/watches, executive folders, desk accessories, balloons, luggage tags, mugs, key chains, notepads, t-shirts, tote bags, ties, napkins, calculators, etc.
- On Anything And Everything You Can Imagine!

Perception Is Everything!

In marketing, perception is everything. Sharing an award logo reinforces the perception among your clients and prospects that your company is a leader in the contact center/CRM industry. It tells them that you are a member of an exclusive group of companies with a record of achievement worthy of high honors and recognition.

Some Guidelines for Effective Use of Awards Logos

Following are some ideas on how to market your awards efficiently, to gain NEW customers and increase market share:

1. Develop a powerful, benefit-driven overall corporate concept and strategy for marketing and promotion.
2. Develop a consistent and cohesive marketing plan where every part of the plan delivers EXACTLY the SAME MESSAGE.
3. Saturate the print, broadcast and online media with a powerful news release in which the importance of the award is clearly defined and elaborate on the authoritative position of the source.

4. Make sure your press release is original and doesn't use copycat, me-too-style wording that turns off every editor.
5. Develop ads that are benefit- and response-driven where your award logo is LARGE (i.e., 1-square-inch size at a minimum) and prominently displayed.
6. Define the meaning of the award, and use it with your company and with your product logo everywhere.
7. Use integrated marketing and prominently display your award logo on the web, in brochures, at trade shows, on stationary, in outbound and inbound marketing, in specialty promotional items (i.e., mugs, golf shirts, etc.) along with your company name.
8. Use your award logo and your company name in each and every correspondence that leaves your company.
9. Be sure you place the award logo adjacent to your company name on your website and social media sites.
10. Use your award logo in press conferences and at all board and user group presentations such as keynotes and seminar presentations. Use it in analyst meetings and any and all presentations and communications with Wall Street.
11. Always use your award logo EXTRA LARGE in full color. Small logos have no value.
12. Be proud of your achievement and say: Our award says it all!
13. Bombard social media with stories.
14. Encourage employee advocacy. Develop a promotion strategy with incentives to encourage your employees to also share the success of their achievements being part of your team with their family, friends, neighbors, community and social media networks.
15. Last, but not least, use social media ads, rent a targeted e-mail or mailing list and send frequent messages to your target audience to inform them about the award and the fact that you are an industry leader in your field

UNDER EXPOSURE AND OVER EXPOSURE IN MARKETING SO MUCH FOR HUMILITY IN MARKETING

Nadji recently said, "I have always been extremely interested in sports because I feel that they offer not only a healthy mind, but also teach one about the importance of teamwork, determination, achievement, winning, losing and satisfaction. I feel there is a strong correlation between what we learned from sports and our ordinary lives."

Why I Love Sports

"Since childhood, I have always loved sports, not only because a healthy mind is found in a healthy body, but also because we can learn very much about motivation, extraordinary achievements and greatness from sports, benefits that are transferable to our lives and our business."

What I Learned from Following the Greatest Achievers in the World of Sports

"One of the most interesting and informative programs that I watch on TV is the annual Pro Football Hall of Fame program, which takes place in Canton, Ohio, and is televised regularly. To me, the Hall of Fame induction represents being the best of the best of the best in sports or any other endeavor. I, therefore, feel it is extremely important to learn how the Hall of Fame inductees got to that point: why they are one in a million and have earned such high esteem among their peers. By learning what they have done, we can always apply the same principles in our own lives to help us achieve the highest possible position, in our businesses, in our jobs and in everything else."

About Humility and Its Significant Place in All Sports, Religions and Our Lives

The authors researched to find a good definition for humility. According to Wikipedia, on the subject of humility, we found the following descriptions:

"Humility is a quality characteristic ascribed to a person who is considered to be humble. Humility is derived from the Latin word' humilis 'which means low, humble, from earth. A humble person is generally thought to be unpretentious and modest: someone who does not think that he or she is better or more important than others."

In other research, we found that humility is described as "a sign of godly strength and purpose, not weakness." Having said all of that, it seems that society does not often reward the humble person. An important case in point is that of Gene Hickerson and his Hall of Fame story.

How Gene Hickerson Nearly Missed Being Inducted Into The Hall Of Fame Because Of His Extreme Humility

By all standards, Gene Hickerson was one of the best, if not the best, offensive linemen in all of pro football. According to documents from the Pro Football Hall of Fame in Canton, Ohio, he was regarded as one of the best, if not the finest, linemen in the Southeastern Conference during the end of his collegiate career.

Hickerson was drafted by the Cleveland Browns in the 1957 NFL draft in the seventh round. He quickly went from delivering plays to the huddle to establishing himself as the most valuable lead blocker for three future Hall of Fame running backs including Jim Brown, Bobby Mitchell and Leroy Kelly.

In plain English, Gene Hickerson paved the way for the above three running backs to achieve extraordinary performance and, in fact, all three were eventually inducted into the Hall of Fame nearly 25 years ago thanks to the great blocking of Gene Hickerson.

Yet, because of his extreme humility, Gene Hickerson was neglected, overlooked, ignored or forgotten by the Hall of Fame committees because he was an extremely humble man.

During the Hall of Fame induction program in the first week of August 2007, his colleagues stated that Gene was so humble, when people asked him, "What do you do?" He would say, "I am an astronaut, a teacher or a mailman or something like that." He always kept a low profile with his loveable, down-to-earth personality.

Of course, we don't know exactly why he was neglected for so many years, but in the absence of that information, if we were to guess, we would say it is because he avoided exposure or marketing his achievements that he did not enter the Hall of Fame until this year, over 25 years later than the three running backs he was instrumental in sending to the Hall of Fame.

It was extremely sad to watch Gene Hickerson receive his Hall of Fame award while in a wheelchair when he should have been inducted to the Hall of Fame more than 25 years previously, preferably ahead of the other three Hall of Famers. That's simply because without his legendary downhill blocking ability, it would be highly unlikely for the other Hall of Famers to have gotten theirs so early in their lives.

Humility in Marketing

Applying humility in marketing will also have near-disastrous consequences. As we have stated many times, if you don't market, you don't exist. In this case, and in our opinion, had Gene been more visible, just as other players were, no doubt he would have been in the Hall of Fame much earlier. In business, too much humility in marketing will cost businesses millions of dollars, and there is no way to remedy the problem.

The Jack Youngblood Story Teaches Us about Courage, Persistence, Determination and Hard Work beyond Imagination

A few years ago, during the Hall of Fame ceremony, the legendary Merlin Olsen, who was one of the teammates of Jack Youngblood, introduced Youngblood by saying that during the first practice in the rookie year of Jack Youngblood, at the end of practice the coach said, "Youngblood, you are the worst football player I have ever seen."

Obviously, as a No. 1 draft choice of the Los Angeles Rams in 1971 as a defensive end, he was expected to perform to much higher standards than he did in his first practice. Youngblood did not take these words lightly, and he decided to work very hard in the subsequent 201 consecutive games (which was a record) to prove to the coach that he was, in fact, better than the worst player the coach had ever seen. Youngblood suffered a fractured left fibula in the 1979 first-round playoffs, but played every defensive down in the title game, Super Bowl XIV, played in five NFC Championship games, All-Pro five times, All-NFC seven times.

The bottom line is that he was most famous for playing the entire 1979 to 1980 playoffs (including the 1980 Super Bowl) with a fractured left fibula – a broken leg! At the end of the game, the same couch said, "Youngblood, you are the greatest football player I have ever seen!" Hopefully you will agree with us that the success stories of these Hall of Famers give us a tremendous amount of inspiration and motivation for achieving higher and higher objectives.

Muhammad Ali's Success Story

As we all know, Muhammad Ali was anything but humble. Yes, he had plenty of talent, but because he was always thinking outside the box, which made him seen abnormal or weird to the public, he always stood out. Indeed, we all remember his rope-a-dope technique or his "Float like a butterfly, sting like a bee" saying, not to mention admiring himself, proclaiming that, "I am the greatest, I am beautiful and my opponent is ugly."

Indeed, he was talented, but this type of aggressive behavior made him stand out and got him, perhaps, more fame and admiration than any other sports figure to that point in history. Throughout his life and most recently after his death, the headlines from USA Today to Time magazine simply read, "The Greatest, Muhammad Ali, 1942-2016."

Clearly, this kind of publicity went a long, long way.

When visiting the famous Madame Tussauds Wax Museum in London, you will notice statues of all of the world's leaders, including all of the kings and queens and prime ministers of the past. Included with them was a statue of only one athlete: that of Muhammad Ali and no one else. Had he been as humble as Gene Hickerson, he would not have achieved 10 percent of the worldwide renown he got.

Donald Trump – Marketer in Chief

Donald Trump's book, No Such Thing as Over-Exposure, is not focusing on humility but on master marketing, and he is, indeed, a master marketer and his accomplishments are exemplary.

Trump turned the 2016 U.S. presidential election upside down with his non-traditional campaign approach. He was campaigning to be commander in chief, but he certainly earned the title of marketer in chief.

Tim Calkins, a branding and marketing expert at Northwestern University's Kellogg School of Management, thinks Donald Trump has been pretty darn brilliant. He sold himself in ways that lured voters, discomfitted the GOP hierarchy and confirmed the tenets of professors who know marketing.

"The whole thing is such a very interesting marketing story and fascinating to watch from that lens. Politics is always about positions but, at its core, about marketing," says Calkins, who teaches MBA candidates and executives.

All the presidential candidates were seeking to build their brand and create one that people would support because they believed it was in their self-interest. But Trump managed to do exactly what great brands need to do, according to Calkins.

First, he was able to generate a huge amount of attention. "That is the first thing you need to do to build a brand – to get people to notice and not to fade away into the clutter that exists today," says Calkins.

Second, he differentiated himself through positions and comments unlike those of any rival. It's involved a distinct nerve – others would characterize it less charitably – to say what others just wouldn't.

Third, he provided a clear and simple perceived benefit. Yes, he takes "unique positions" but he ties them to the benefit "of making American great again." He puts it all together in a pitch that resonates with many people. Clearly there is no humble pie being served in the Trump Tower dining room.

Many of us have mixed feelings about the subjects of humility and boastfulness. Is it better to be humble and be respected by your colleagues, or should you take a lesson from some people who reside on the other side of humility and endeavor to make yourself world famous? The choice is yours.

In marketing, however, there is no room for humility, as we can learn from Mr. Trump. You must dominate the media to dominate your market.

Nadji Tehrani's Participation and Modest Achievements in Sports

In his teenage years, Nadji was an average soccer player and an accomplished cyclist. As a cyclist, he came in second in a six-mile race among top high schools in Tehran.

However, as a Little League soccer coach of children ages 8 to 12, his team won the city championship in 1977 in Stamford, Conn. They played 45 games that year and their record was 45-0. The average score was 18-0, and his team allowed only two goals all year. If you are familiar with soccer, you know that such a good record is unheard of in that sport. Nadji often comments that you might say he was a better coach than a player. Frankly, there is a strong correlation between coaching a sports team and managing a company.

Award Winners Differentiate Themselves In A Powerful Way But...

Awards without marketing are practically worthless. And, we have also stated that quality and marketing are not part-time jobs. Indeed, we are aware of companies that are award winners, and they are extremely high quality. By doing practically everything right, including award marketing, these companies are industry leaders and are growing and becoming more profitable each and every year. We think there is a good lesson to be learned from such companies.

Indeed there is no more powerful differentiation tool than winning a coveted award from a prestigious and highly respected publication of the industry.

Now That Most Important Question:
WHAT MAKES AWARD WINNERS SUCCESSFUL?

In our opinion, it is great to congratulate them, which we do sincerely. It is great for them to market themselves and make the most of their awards and differentiate themselves from competition. But, the underlying question is: How do they do it?

Every CEO and every savvy manager would like to know how some of these companies are winning awards every single year.

Nadji stated, "For example, the company InfoCision Management Corp., has won consecutive MVP Quality Awards every single year since the inception of the award.

"To me, they are a model company and exemplary in every way. Frankly, I have been studying this company for several years, and I speak from experience. Not only do they have relentless and demanding quality standards, but they also conduct award marketing better than anyone. When you do both of the above well, there is no way to avoid prosperity.

"I have always felt that teleservices (call center outsourcing companies) are the most knowledgeable people in the contact center industry. The reason: they only do inbound, outbound and CRM work for corporate America and, as such, their experience far exceeds that of anyone else, bar none."

For the above reasons, we decided to study the philosophies of some of the teleservices companies that have won MVP Quality Awards. The objective was to find out what they do differently and what we can all learn from these companies. In search of the answer to the above questions, we found the following statements that were made by senior management of these teleservices companies extremely helpful. The underlying philosophies that make them so successful are presented here.

And now we would like to share some of them with you.

"We believe that every customer contact must meet or exceed the standard we establish for quality performance. Our motto is, 'We are only as good as our weakest link…so even our weakest link must be strong.' We don't look at group averages as indicators of success. Instead we measure our success by the success of each individual."

"Our motto is to treat each customer as if it was your parent, spouse, or child stuck on the side of the road and treat each call as if it is the only call you will take today. We teach our associates to be excessive about delighting the customer."

"Our standard quality requirements include a 130-point review on each call monitored. Our stringent quality standards ensure our ability to outperform the competition."

"To be truly effective, an outsourcer's quality assurance program must do more than simply ensure that calls are processed correctly. It must be a total commitment to quality that encompasses all aspects of the business, including recruitment and hiring, contact processing, workforce management, performance reporting and administrative support."

"One of the most unique aspects of our company is its entrepreneurial spirit. We promote contact center employment as a career and not as a temporary job. We focus on recruiting from within."

"Our philosophy is simple: Selling is customer service, and the only good sale is a quality sale. Outsourced clients are offered the best opportunity to exceed sales projections and achieve the highest quality sales possible, while they solidify relationships with customers entrusted to us."

"Early on, we concluded that growth alone, including new call centers, expensive technology, numerous workstations and thousands of employees, is not the key to teleservices success and long-term client relationships. Quality is."

"As consumers have become more accustomed to contact centers and more demanding of the service and support they receive, it has become less important to focus on call mechanics and metrics, and more important to focus on customer satisfaction and the overall customer experience."

"Today, our definition of quality has evolved more fully: quality is an ever-evolving perception carried by the people involved with and impacted by the services delivered by our company."

"Our home agent model lends itself to lower costs. In the past, many companies selected offshore companies to reduce costs. Unfortunately, the tradeoff was that many customers lacked a language and cultural affinity with the representatives of these companies."

"We believe the most solid outsourcing relationships are those in which the client allows us to truly embrace their culture, and in turn, treat us (as an outsourcer) as an extension of their company. We strongly encourage this and recommend active day-to-day involvement in the operation of the program."

Industry Blunders

No matter how good your marketing is, an industry blunder can realistically sabotage any expectation you may have from your marketing.

Return to Sender

Long ago UPS decided to outsource a significant portion of its back office to conduct a significant amount of telemarketing. UPS asked leading agencies to provide a proposal outlining guidelines, costs and expected results from the proposed campaigns. One company spent huge amounts of resources to develop a meaningful proposal for this huge application.

The Advent of Telemarketing®

How Nadji Tehrani started the Telemarketing® industry and how he obtained the registered trademark for the word, "Telemarketing®"

Nadji and Steve photos:

Nadji states, "People want to know how I started Telemarketing® magazine and subsequently the most successful and longest running industry trade shows."

"In 1984, we decided to include a major amount of training articles in Telemarketing®

It gave the proposal to a secretary who inadvertently mailed the proposal to the president of UPS using FedEx to deliver the package. Obviously as large as UPS is, at the very least it would expect any proposal would be sent to the company via its own UPS shipping services, and certainly not using its competition, FedEx. This huge blunder killed the possibility of a truly major account to be acquired by the agency.

Filing Sales Leads Is Never Acceptable

Another contact center company advertised heavily in TMC's CUSTOMER magazine for approximately one year. At the end of the year, the marketing vice president called TMC and explained it had spent a significant amount of money to create qualified sales leads through the advertisements and it was quite concerned that its salespeople had not received a single lead of any kind as a result. TMC actually found this very hard to believe. A TMC editor asked the advertiser to please review the process of lead turnaround.

The following month, the vice president called back and asked to speak directly with then CEO Nadji Tehrani. He said, "Nadji you will be happy to note I have terminated the secretary of our marketing department. Nadji asked him why. The vice president said, "She was simply filing the leads and not sharing them with the salespeople. This type of blunder is disastrous."

We must communicate the marketing plan details carefully across the various department team members to ensure they understand their role and how to create sales opportunities and customer relationships with our company.

Filing sales leads is never acceptable. The bottom line is a company must pay attention to details, as one mistake can cost the organization substantial dollars.

May we suggest that all companies reading this book also buy a copy of Taking Your Customer Care to the Next Level™, written by the same authors, and available on Amazon.com in hardcover, paperback and Kindle e-book versions.

magazine as well as seminars at trade shows. I want to say in fact, Steve Brubaker of InfoCision Management Corp., came to the trade show and met me personally in 1986."

"Steve understood Telemarketing® was a growing industry and asked to learn and be introduced to the leading minds of the industry. Those who were movers and shakers."

"So, as the industry began to grow, many countries including the U.S., Japan, Korea and

Europe started telemarketing training universities. The Japanese opened a new university and invited me to attend the grand opening ribbon-cutting ceremonies. In fact, most of the developed countries were sending from 20 to 40 people each year to our events to learn and then return to their countries. As such we contributed to the development and globalization of call centers."

***

Chapter 5

The Basics of Effective Trade Show Marketing

Effective Exhibiting; Easier Said Than Done

Creating a successful trade show exhibit begins long before the convention is held. Before even considering exhibiting, you will need to keep in mind that no one will buy anything of value from you if they have never heard of you and your product. To create a proper perception of your company and its products/services, you need to embark on an effective advertising campaign long before exhibiting. Advertising will properly position our products or services for purchasing decisions.

Don't Confuse Lead Generation with Selling

One of the major benefits of trade show exhibiting is lead generation. Many business people expect to always sell on the show floor, or shortly thereafter. While this may happen in a few shows, by and large the main goal of exhibiting should be to generate sales leads. Subsequent to the show, these leads must be placed in the hands of competent salespeople to be converted from prospects to customers. The key is to ensure that you employ a truly effective sales force that will, in fact, follow-up.

To survive in today's economy, your business cannot afford to waste worthwhile sales leads. The basis behind trade show leads is that an executive attendee usually spends between $1,500 to $2,000 to attend a trade show of interest. As such, no one in the world is more qualified to buy your product than a trade show attendee, provided you are exhibiting at properly targeted shows.

In summary, exhibiting is one of your most effective marketing tools, provided it is done correctly with appropriate preshow marketing, during-show marketing and post-show marketing. If you are not prepared to exhibit the right way, you should not exhibit at all.

The Extinction of the Trade Show Dinosaurs

As anyone who has attended trade shows over the past few years must surely have noticed, there has been an undeniable drop off in attendance across the board, but most particularly at the huge, everything-under-one-roof horizontal shows that try to offer everything for everybody.

There are a number of factors that have led to this. The slowing economy and declining stock prices have led to tighter budgets, resulting in cutbacks in business travel and rethinking of how companies should spend their precious resources and what they should receive for them. We have witnessed a paradigm shift from huge, super horizontal shows to small, regional, tightly focused vertical shows.

Avoid The Illusion of Crowded Horizontal Shows

One reason large shows are outdated dinosaurs headed toward extinction is because they have become too expensive for their exhibitors. Companies are no longer making decisions to exhibit at or attend events just because they always have. Events today have to show their value each and every event. There are those who are, mistakenly, of the opinion that the larger the crowd, the more sales leads you get from a trade show.

While a company may get more sales leads from a large show, say 1,000, compared to perhaps 500 sales leads from a smaller, more focused show, often fewer than five percent of the attendees from a large show are qualified leads. Very few who have the authority to purchase may be interested in your product or service.

On the other hand, one can almost guarantee that the leads at the more highly focused vertical shows will be more highly qualified because marketers have become much savvier at attracting more highly qualified attendees to smaller vertical shows. Also, attendees prefer the more intimate settings of small shows where they can easily find the vendors they are looking for as compared to trooping around a huge hall or from hall to hall at large shows searching for a particular vendor. In addition, a smaller show provides better networking opportunities.

Let us say that, for example, it costs the exhibitor $8,500 in total to exhibit at a smaller show to get 500 leads. The total cost per highly qualified lead is $17. Contrast this to the $150,000 a company may spend exhibiting at a large show to get 1,000 leads, which means it spent $150 per partially qualified lead. The exhibitor will then have to spend more time and money to eliminate the 95 percent of unqualified leads. In short, if we assume that only 5 percent of the leads are qualified at a large show, you actually only get 50 leads! In that case, the real cost per qualified lead is $300, plus additional qualification charges of $2,000.

To do a proper return on investment analysis, we must add the above lead qualification costs to the exorbitant cost of exhibiting at larger horizontal shows. If you add up the above, the total cost of qualified leads generated by large horizontal shows you will be comparing about $350 per lead at large shows vs. only $17 at smaller, more focused vertical shows. When you look at the two, there is no comparison.

Superior CRM Is Only Possible At Smaller, Focused Shows

A small, more focused vertical show brings in anywhere from 1,000 to 2,000 attendees (provided the show sponsor and exhibitors do a proper job of team marketing). The exhibit hall itself is relatively small, so that every attendee has time to stop by every exhibitor's booth and have an adequate amount of time to discuss the customer's needs and solutions the exhibitor has to fill those needs.

Consequently this type of proper dialog between the exhibitor and the attendee is far more conducive to closing the sale as opposed to the one to two minutes the exhibitor and attendee have at a large show with a small fraction of the total attendees and where nothing useful will come out of it.

At small, focused shows, vendors have a great opportunity and the time to develop a real relationship with existing customers as well as gain new ones.

Part of the illusion of the value of large shows was the prominence of large booths. The last year has proven that it's not the size of your booth that defines your company anymore it's the product in the booth. Shows have become what they were designed to be, a forum for selling products. Although branding is important, it has become less of the dominating decision-making factor in doing a show.

The Bottom Line

The bottom line is that every attendee at a smaller show is far more qualified. Although a smaller show may bring in a smaller number of attendees, the exhibitor will gain far more sales from such attendees, not only because the attendees are pre-qualified, but also because, in this economy, corporations are more likely to send decision makers to shows.

There is a myth in the industry that the more leads you get, the more successful the show. That line of thinking no longer has any validity in the new paradigm of trade show marketing. Recent cancellations of many mega shows add further credibility to the comments made in this editorial. When you take a close look, there is no comparison between the outmoded dinosaur that is the large, horizontal trade show and the nimble, more cost-effective medium that is the new, relationship-building vertical trade show.

The Best Kept Secret in Trade Show Marketing:
They Don't Teach You *This* At Harvard Business School!

Every year Technology Marketing Corporation participates in a dozen trade shows, conferences and user-group meetings, in addition to taking eight or 10 specialized trips, to bring our valued readers the absolute finest in state-of-the-art marketing.

At one of these conventions, TMC decided to have a low-key presence. We rented a booth and decided to hire a local booth attendant, who we'll call Type A, to distribute *Telemarketing*® magazine and collect sales leads.

On the first day of the show, the hired booth attendant was completely trained by one of TMC's sales managers as to exactly what was expected of her. When the exhibit hall opened, the sales manager became increasingly concerned about the attendant's lack of motivation, enthusiasm and interest. She had a laid-back attitude and stayed as far away from the aisles as she could, as though she wanted to avoid the attendees.

Her attitude was that the world owes her a living, the magazines are on the table and if people want them, they will take them, and if people *insist*, she'll run their identification cards through the imprinter so they can get on the mailing list! She seemed far more interested in talking to the young sales men in the adjacent booths about social matters, lunch appointments or which hospitality suite she would go to!

If a customer wanted to buy a book or order a subscription, he or she literally had to interrupt the social talk to get more information. The sales manager grew increasingly impatient with her performance and gave her repeated warnings throughout the first day. Due to this attitude, she collected only 29 sales leads during the first day.

Nadji arrived at the convention the second day of the show and learned of the booth attendant's unacceptable performance. He immediately had a serious talk with her, explaining that TMC had spent several thousand dollars to come to the show to generate new business, not to socialize or pay her $25 an hour so she could talk to non-customers and set up lunch, dinner or other appointments. Since Nadji noticed some talent in her, he decided to give her one last chance to shape up or ship out!

The bottom line: she did not respond and she got a total of only 58 sales leads for the first two days of the show! She was dismissed immediately.

Nadji then registered a very strong complaint with the temporary agency and insisted he receive the agency's best available booth attendant or he would not pay them a dime and would bring the matter to the attention of the local Better Business Bureau. His justifiably bitter complaint paid off handsomely.

The next day an outstanding booth attendant, who we'll call Type B, was sent to the TMC booth. Unbeknownst to Nadji, she arrived in the exhibit hall (somehow) two hours before the exhibits opened and she immediately read every single magazine, brochure, flyer and book she could get her hands on. She developed (on her own initiative and without any training whatsoever) a remarkably accurate, convincing and memorized script that she presented to virtually everyone who passed by the TMC booth.

Due to inadequate transportation availability that day, TMC staff along with Nadji arrived nearly two hours after the show had opened. Naturally concerned about the status of the new booth attendant, Nadji rushed to meet her.

What a pleasant surprise! He could not believe his eyes. Nadji stood speechless right in front of her, wondering when she had learned so much about TMC's magazines, books, digital services and conferences. Since she did not know Nadji, he asked her several questions and she responded with unbelievable accuracy! She handled the attendees with such professionalism and care, the likes of which he had never seen in his life. Her warm, sincere and bubbly personality fueled by her positive, can-do-attitude literally stopped every attendee to hear the compelling TMC story!

As for results, she was in a class by herself. Although the last two days of any show have much slower traffic than the first two days, her accomplishments were sensational! She distributed a record 3,000 magazines, calling for another 1,000 emergency shipment. She sold dozens of books and got more than 400 sales leads vs. only 58 for Type A.

The evidence is black and white! A positive, can-do-attitude, self-motivation and initiative generate results. Without those attributes, there is no point in getting out of bed in the morning!

There are many lessons to learn from this true story, among them:

1. Do not take people with a Type A attitude to the show.

2. Before you blame the trade show for poor sales lead generation, take a hard look at your staff. Are they Type A or Type B employees?
3. Before you take employees to a trade show, have them read this book first. Then make it crystal clear to them that they are not going to the show to socialize, they are going to work hard (12 hours straight a day) and generate business. If they still want to go, perhaps you are fortunate enough to have Type B staff members.
4. Also, take care of your Type B employees. Nadji paid the Type B attendant a healthy bonus, gave her his business card and told her that anytime she wanted to work for TMC full time, to give him a call.

There's much more to trade show marketing than just showing up! We wonder, do they really teach this sort of thing at Harvard Business School or is this not academic enough?

Marketing Service Agencies – Why Partnering With Experts Pays Off

When corporate executives begin to see the value of an integrated multi-channel marketing strategy, usually their first inclination is to develop an in-house solution. If they are open minded, they will first read every piece of literature ever written on the subject, and then attend the major industry conferences and trade shows and talk to colleagues with similar interests.

They will also hire a seemingly knowledgeable consultant, as well as buy the right hardware and software, trusting the technology will be applicable to their end use, as well as compatible to any existing equipment.

Furthermore, they might find a training company to come in and train their people. And when all is said and done, they will hope and pray that everything will work together.

If (and that's a big if) all of this comes together, each of these companies will need an unbiased marketer to put together a sales package that will project the company's products and services successfully. The difficult-to-come-by, unbiased marketer will then have to test a variety of packaged offers to determine the following:

- Is the product in fact marketable?
- Is the price and performance competitive?
- What marketing program will be best received by customers?
- What is the ROI? Will the marketing campaigns be profitable?

Once these criteria are established, the marketer will have to develop several campaigns and test their effectiveness, then devise and evaluate the back-end processes to respond to inquiries and close the sales leads.

If you take a hard look at this scenario, you will soon discover that the chance of everything coming together is very close to nil and the chance of making at least one or two mistakes each step of the way is better than 60 percent.

At this point these executives may have invested anywhere from $150,000 to $1,000,000 or more (depending on the size of the operation) with no assurance that they have a viable operation. They'll feel as though they have reinvented the wheel without knowing if it will work. In addition, others in top management and/or stockholders are constantly applying pressure for return on all investments. They will undoubtedly say to themselves: If I had to do it over, I would have done many things differently. Of course, hindsight is 20/20!

BUT you don't have to do it that way. There is no need to reinvent the wheel and spend a lot of time and money to wonder if it will all be worthwhile. There is no need for the costly hindsight. There is indeed a far more attractive management alternative – hiring a reputable marketing service agency!

Service agencies have traveled this road many times before. They have made all these mistakes, paid their dues and are willing to share their valuable experience with you for a fraction of what it would cost you to do it yourself!

What makes most marketing service agencies unique is that they market virtually every type of product and service and thus have acquired that unique savvy that spells the difference between success and failure in marketing. That's the kind of experience you gain by being on the firing line for a long time! No school, university or textbook can teach you that!

A reputable service agency is that unbiased marketer that can tell you, after a small pilot project, whether your product or service is marketable or if it's packaged right and, if not, what you have to do to solve the problem! Best of all, the cost of this pilot test project is usually under $10,000!

If you compare the two options described, it becomes abundantly clear that using a service agency is by far the more desirable option. This is not to say, however, that one should never consider building internal operations for customer acquisition and retention. What we are suggesting is that using a marketing service agency can help you minimize risks, avoid costly mistakes and train your staff for any potential future in-house cutover, all at a more reasonable cost.

Savvy marketers know the value of not putting all of their eggs in one basket. That is, in the event that you follow the suggested scenario of getting started with a service agency, it is not advisable to bring all of your marketing in-house. Many companies, particularly those with inexperienced management, fail because they never think of or even prepare for the all-important business continuity!

Fire, vandalism, strikes, hurricanes, equipment failure, blackouts and computer hacks can, in fact, bring your marketing operation to a halt for a long time. These are real and important reasons why astute marketing executives must think of continuity and keep one or more service agencies actively involved in their marketing operations.

Successful Marketing at a Trade Show Is A Two-Way Street

As the global competition intensifies and as recessionary trends continue, it becomes increasingly indispensable for management to reduce costs and maximize return on investment. It is fair and factual to say that in sales and marketing, nothing offers the maximum ROI for B2B sales like trade show marketing combined with digital marketing lead generation and follow-up. The synergistic power of the two marketing methods is simply incredible and beyond one's imagination.

Background

With the exception of digital marketing, trade show marketing is, by far, the most effective way to market to business customers. However, trade show marketing, like anything else, is successful if, and only if, it is conducted properly.

There is a gross misconception among some exhibitors of shows that if they collect a lot of leads, qualified or not, or if there is heavy floor traffic, then that is a good show!

Astute marketers know that the process of lead qualification is extremely costly and time-consuming, particularly when only a few percent of the leads are qualified!

Experienced exhibitors also know that you don't go to the fireplace and say: "Give me heat...then I will add wood!"

They know that successful trade show marketing is a two-way street. The trade show sponsor must perform a gamut of marketing functions to bring qualified attendees to the show floor. It is the exhibitor's job to also follow several vitally important pre-show marketing steps in order to give the attendee a reason to enter the exhibitor's booth!

Checklist for Successful Trade Show Participation:

For successful exhibiting, ask yourself these 15 vital questions before you exhibit at any trade show:

1. Do your booth graphics grab attention quickly?
2. Do your graphics communicate in a few seconds what type of product or service you offer?
3. Do your graphics give the attendees a good reason to stop by and examine your products or services?

4. Do your booth design and graphics communicate a benefit for any attendee to stop by your booth?
5. Have you advertised in the leading industry publication(s) inviting readers to visit your booth? Did you offer them a FREE VIP Pass to do so?
6. Have you called all of your top 100 customers and invited them to attend the show and visit your booth with complimentary VIP PASSES as your guest?
7. Does your company have name recognition? For example, would the attendee know what you sell by simply seeing your company's name? (i.e., Coca Cola sells soft drinks; McDonald's sells hamburgers, etc.)
8. Have you taken the right people who are well trained in exhibit marketing to the show with you? Remember, this is vitally important.
9. Have you sent several mailings (print and digital) to your database, each time giving them a new, important reason why they must visit your booth?
10. Are you visible everywhere as a company?
11. Do you regularly get your message across through your advertising?
12. Do you come up with a major new attraction or attention-grabbing idea in your booth to make YOU STAND ABOVE THE CROWD?
13. Do you sponsor events at the show to draw all delegates' attention to your booth?
14. Does your booth staff have proper boothmanship? Are they sitting around talking among themselves, reading newspapers, and smoking, eating or drinking or talking to the office instead of aggressively seeking out customers from the aisles 100 percent of the time?
15. Last but not least, is there anything in (or about) your booth that would encourage a potential customer to come to your booth instead of (or in addition to) your competitors?

If your answer is positive to most or all of the above questions (and we certainly hope it is), chances are, you will have a very successful trade show. If it is not, do not expect to have a successful show, and when that happens, don't blame the show sponsor for your own failure for not giving the attendees a good reason (or any reason) to visit your booth.

Be sure to mail VIP invitations to your entire database (or at the very least, to your key prospects) inviting them to visit your booth to learn about your existing or new products and services. Remember that at least one-third of the traffic in your booth comes from your company's own direct marketing activities PRIOR to the show!

Be sure that your booth graphics tell all attendees (in less than two seconds) what you do or offer. Why? Given that attendees are senior executives from major, growing corporations, they'll only have two seconds to decide whether or not to come into your booth. They don't have time to stand there trying to figure out what your company offers.

For example, some suggested sign copy might read as follows:

"We Are A Top Marketing Service Agency. May We Help You?"
"_______ Industry Software Is Our Business"
"The Leading _______ Industry Vendor."

If the name of your company does not explain what your company does, it is vitally important that your booth sign provides this message. In all cases, large block letters supported with relevant well-lit graphics with pleasing colors (i.e., yellows, warm reddish orange accented with a touch of magenta, aqua green and royal or navy blues) will encourage attendees to enter your booth. Otherwise, if they have to spend a lot of time to figure out what you do, while staring at an unattractive booth, you can bet they will walk by.

Remember, you and 100 other exhibitors are competing for the attendees' attention. The better job of marketing you do in advance, the more traffic you will see at your booth. This is simply the logic behind trade show marketing.

That's why the companies that are aggressive marketers will always have a huge number of attendees at their booths throughout the entire duration of the show, regardless of attendance. On the contrary, those companies that do nothing to promote their participation in advance, and do not comply with any of the above, will not have much traffic.

It stands to reason that if attendees don't know who you are or what your company is all about, they have no reason to stop at your booth. Consequently, what you do in terms of aggressive marketing and advertising prior to the show is the key to success in exhibiting at any trade show.

In short, like anything else in life, when it comes to a trade show, you get exactly what you put into it.

The Plain Facts about Trade Show Marketing

The Trade Show Bureau, a leading trade show educational organization, has conducted research on trade show exhibiting. The following are some of the bureau's findings.

A. **Trade shows draw quality audiences.**
   According to the bureau's research, 86 percent of trade-show visitors have buying influence, and 59 percent plan to purchase within a year. Only 17 percent have previously been called on by an exhibitor's salesperson.

B. **Trade-show visitors spend quality time on the exhibit floor.**
   The bureau's findings show that over a two-day period, more than one-third of trade show visitors spend more than eight hours at exhibits, with the average visitor spending 21 quality minutes at each of 17 exhibits.

C. **Trade shows influence sales.**
   According to the bureau, at one trade show, 90 percent of the resellers bought one or more of the product types on display within nine months after the show, and 91 percent planned additional purchases in the following 12 months.

D. **New products and developments attract people to trade shows.**
The bureau's research shows that 50 percent of trade-show visitors attend trade shows to see new products and developments.

E. **Trade-show visitors hold top positions.**
Almost one-third (29 percent) of trade-show visitors hold top management positions – owners, partners and presidents.

F. **Trade-show visitors have buying power.**
The majority (86 percent) of trade-show attendees have buying power. Eighty three percent know exactly what they want to purchase.

G. **Trade shows reach prospects for less than the cost of sales calls.**
According to the bureau's research, in 1987, the exhibit cost per visitor at trade shows was slightly more than half the cost of a sales call.

H. **Trade shows reach unknown prospects.**
Eighty-three percent of trade-show attendees have not been visited by exhibitor's salespeople.

I. **Exhibitors who place six full-page ads pull 56 percent more visitors than non-advertising exhibitors.**

J. **Booth personnel should be knowledgeable, friendly and approachable – not aggressive.** Insufficient product knowledge is the major complaint among attendees asked to rate booth personnel. Another common complaint is overly aggressive salespeople.

K. **Trade show leads reduce sales calls and closing costs.**
The research shows that the average number of follow-up sales calls needed to close a qualified trade show lead is only 0.8 and closing costs are reduced by almost 75 percent over closing costs without leads.

L. **Larger exhibit sizes generate more traffic.**
The bureau's research shows that the greater an exhibit's size, the higher the percentage of show traffic that stops to visit.

M. **At least 33 percent of the visitors to a booth were there as a result of pre-show promotion.**

N. **More than 50 percent of trade-show leads don't require a sales call to close.**
For 54 percent of the orders placed after a trade show, a personal visit by a salesperson was not required.

Positively nothing produces a return on investment from your B2B marketing dollars like trade show marketing!

Pre-Show Marketing

One of the least understood parts of successful exhibiting is pre-show marketing. Millions of dollars are lost by inexperienced marketers who have a misconceived idea that they can just show up out of nowhere with an uncreative booth and sell lots of products. When they fail, they blame it all on the trade show in which they exhibited. Nothing could be further from the truth. To find the real source of failed exhibits, all they have to do is look at themselves.

Exhibiting, like anything else in life, operates on the principle that the more you put in it, the more you will get out of it. This is especially true of pre-show marketing. If you do nothing in the way of pre-show marketing, you'll gain very little from your exhibiting efforts.

Creative Contests Encourage and Increase Booth Traffic

One of the most effective ways to increase booth traffic at the show is to set up raffles and contests where the drawing for a prize, such as a trip to an island resort, will be made at your booth.

You may also want to require everyone entering the contest to fill out a questionnaire so you can do post-show follow-up marketing to qualified prospects.

Examples of Trade Show Pre-Show Marketing Strategies

Here are two real-life examples of well-planned pre-show marketing strategies which were found to be extremely effective in increasing booth traffic for the individual companies involved.

Red Lobster

One exhibitor to a TMC trade show developed a strategy to draw attendees to its booth by highlighting the fact that its company is based in Maine, in which lobster is a well-known symbol. The company displayed an enlarged photograph of a red lobster that attracted many attendees, as most people are interested in lobster and enjoy lobster delicacies. The red color of the lobster display helped draw attendees to the exhibit booth.

The goal of the exhibitor was to encourage call centers to expand operations in Maine. The exhibitor was the economic development organization for the State of Maine, and its promotion was very effective.

The moral of this story: Come up with an idea and use colors and symbols that reinforce your objective.

What's Missing?

In this example, the company designed a marketing campaign to encourage attendees to visit its exhibit booth to see if it had the missing piece to a puzzle, which included a chance to win a great prize.

The company displayed a giant picture puzzle map of the U.S. with one of the states missing. It designed the map to fit on the wall of the exhibit show floor to attract attention. But it left one state missing.

Random pieces of the puzzle were mailed to high-quality prospects on its list who were likely conference attendees of the show. It knew which conference attendees were most qualified as those who paid hundreds of dollars to attend.

The company promised that individuals with the winning piece of the puzzle would receive a trip for two to Hawaii with all expenses paid. The excitement generated a large number of attendees to its booth checking in to see if they indeed had the missing piece; this provided the company with a healthy number of qualified sales leads.

Don't Let Them Go Cold – Include a Plan for Quick Response

It makes no sense to invest significant resources in generating qualified prospects if you aren't prepared to strike while the iron is hot.

Your trade show marketing strategy must always include a follow-up communication process for cultivating the lead and converting them to a customer of your organization.

Immediately upon returning from the trade show, you must assign the appropriate team member to follow up directly with the prospect and verify the nature of their requirements to further qualify the customer lead. Ensure your plan is adaptable so each sales lead properly fits into the sales cycle and is prioritized within your sales organization.

Chapter 6

Permission Marketing – The Secret to Engage High-Level Decision Makers. The 69 Common Sense Rules Of Marketing

As the founder and organizer of the contact center industry since the early 1980s, Nadji Tehrani still owns the trademark to the word Telemarketing®. He was recently quoted as saying, "I have come to the realization that permission marketing is by far the most effective way to sell anything."

Unfortunately as the market share of Telemarketing® increased throughout the 1980s and 1990s, people hired staff and put them on the phone with little or no training. Specifically in the business-to-consumer marketplace, the impression became anyone could be hired as a telemarketer regardless of skill.

Companies began searching for the cheapest agent on the planet, scouring the globe for low-cost labor alternatives, seemingly not giving a thought to perceived communication barriers due to differences in cultural perspectives. Without an emphasis on quality, consumers became outraged by the volume of unsolicited calls they were receiving from an ever-growing number of competing companies. The calls felt intrusive to people in their homes, and what added more salt to their wounds was that some of the salespeople were untrained and inexperienced, even downright offensive.

The outbound telemarketing industry grew so fast that federal and state legislators as well as government watchdog regulators began to take action. Laws were passed limiting the ability, even of legitimate business, to proactively contact potential new customers. Outbound consumer prospecting calls were effectively legislated out of business and are no longer legally permissible without documentation of a business relationship.

On the flip side, business-to-business sales were not regulated to the same extent. Consumer protection laws normally do not extend to business organizations. In any case, the environment is forever changed and all customers today demand

R-E-S-P-E-C-T.

It is vitally important to have permission to market directly to an individual or business.

As CEO of TMC a few years back, Nadji Tehrani was conducting a meeting in his office to discuss business strategy. A phone call came to him from a salesman who asked, "Mr. Tehrani, is this a good time to talk?" Nadji asked for the purpose of the call, and the salesman stated he could save the company at least 40 percent on its printing costs, and would be pleased to call back when Mr. Tehrani would have time to listen to his offer. Of course, the opportunity to save 40 percent was intriguing, and Nadji immediately wanted to hear more.

So Nadji canceled the meeting in his office and asked the salesperson to proceed with his presentation rather than scheduling a later call. The salesperson had studied Nadji's company, TMC, thoroughly and was aware of the significant volume of materials the media and publishing company was contracting for at the time and the amount it was spending on printing. The salesman had done his homework and was prepared to offer the 40 percent savings. Nadji evaluated the proposal carefully and TMC naturally became a customer.

If the salesperson had just called in pitching, Nadji would not have interrupted the meeting and listened to him. It was only because he asked permission to proceed that Nadji was responsive to hearing the pitch and in turn canceled the meeting. The smart marketing strategy of mentioning a 40 percent savings opportunity up front, of course, also caught Nadji's attention and ultimately led to the sale.

The problem is that salespeople rarely ask for permission. You would likely agree when we say that less than 1 percent of sales pitches begin with the most basic courtesy of asking the customer for permission to proceed with the presentation of an offer.

Do You Have a Sales Prevention Department in Your Company?

Most Companies Have One, But They Don't Know They Do

Many companies actually have a sales prevention department, but they are completely unaware of this fact. When we say sales prevention department we don't mean that these companies literally have a separate department with that title. However, the regular violations of certain important rules actually constitute a cancerous problem within many companies. To succeed in business, you need to understand your customers' needs as well as their customers' needs.

Sales and Marketing are Everything in Every Company

As you may know, we have been students of marketing our entire lives. In fact, Nadji developed a marketing test at TMC that 99 percent of the marketing managers who have taken it failed. In our view, the test is a simple one and contains the basic knowledge that every true marketing manager must possess. There is no point in hiring a marketing manager who cannot even define marketing.

In many companies, the sales department is regarded as the most important department in the company. We do not subscribe to this thinking, because we feel every department is equally important. Having said that, our frequent associations with many CEOs within our industry and elsewhere have led us to believe that most companies, in fact, consider the sales department one of the most important, if not *the* most important department.

In our way of thinking, this is not true. We feel that if you are going to rank the departments, marketing should come ahead of the sales department. Here is why.

All Sales Begin with a Sales Lead

Among the paramount responsibilities of the marketing department are to create awareness about the company, articulate the benefits of dealing with the company and highlight the company's differentiation from its competitors. The cumulative results of the above-mentioned marketing functions eventually lead to the all-important lead generation that is vital to any company's growth and prosperity. In other words, the sales department will be crippled if the marketing department does not generate a stream of continuous, qualified leads for the sales department.

Sales Prevention Diagnostics

Having stated the importance of the sales and marketing departments, there are many details that need to be addressed if sales prevention is to be avoided. Here are the areas that are most likely to contribute to sales prevention:

1. **Ignoring the Golden Rule of Integrated Marketing and, Most Importantly, Ignoring the Golden Triangle**

   When a company ignores the rules of integrated marketing and the Golden Triangle, which includes the combination of print, online and event marketing, the company has, in fact, prevented maximum lead generation for the sales department.

2. Ignoring Marketing Completely

   Believe it or not, many companies give lip service to marketing and, as far as we have been able to study, such companies either go under or, if they exist at all, they really don't get anywhere. Nadji recalls a pair of companies that started out in the Chicago area at the same time. Company A was a master marketer and Company B did not care about marketing at all. To make a very long story short, the owner of Company B is still struggling and has gotten nowhere in the same period of time!

3. Wasting Sales Leads

   Many companies spend a tremendous amount of money every year attending trade shows or advertising in print and online and generating a considerable amount of leads. However, research indicates that as many as 70 to 80 percent of sales leads generated are either ignored completely or followed up too late to be of any use. Indeed, this is one of the leading causes of sales prevention.

4. Ignoring your customers' needs and, most importantly, ignoring your customers' customers' needs.

   In this highly competitive business environment, the companies that go beyond the call of duty are those that will survive. Once again, to succeed in business, you need to understand your customers' needs as well as your customer's customers' needs. Remember that Customer Care™ is the only sustainable competitive advantage.

5. Ignoring sales training

   Many companies, particularly the entrepreneurial small and medium-sized companies, have a tendency to ignore sales training. This is practically unthinkable. How can anyone expect a sales person to sell anything without knowing the benefits and features of the products or service they are expected to sell? Believe it or not, this problem continues to exist.

6. Having a Nasty CEO

   Many entrepreneurial companies are started by ego-driven individuals. Wisdom and professionalism are substantially ignored when that CEO is dealing with customers. These types of individuals must never be faced with customers; it takes only one nasty remark to destroy a million-dollar deal. We have observed this situation several times in the past.

7. Having a Loose Cannon on the Sales Staff

   This is, perhaps, the most damaging situation for any company. A loose cannon can create not only a tremendous amount of unnecessary problems, but he or she can destroy a relationship and prevent any and all anticipated sales.

8. Having a Loose Cannon Anywhere in the Company

   Obviously, such a person must not be tolerated by any responsible company. That individual could not only destroy the morale of the employees and create problems, but also, when contacted by customers, can create yet another major problem by destroying your relationship with the customer.

9. Having a Chronic Complainer on Staff with a Bad Attitude

   Some people are never happy unless they are unhappy. They are the individuals who complain chronically. Not only do these people destroy morale within the company, they have the potential to significantly damage the morale and attitude of the sales department. This problem should also not be allowed under any circumstances, at any company.

10. Having an Unwise Sales Compensation Program

    One of the most powerful management tools is to develop a mutually beneficial compensation program that fosters accountability on the part of the sales people.

11. Having a Lack of Teamwork

    As a sports enthusiast and former coach for Little League soccer, Nadji has learned that nothing is more important in any organization or any sports team than teamwork. Nadji states, "One of the greatest things that I have heard along these lines is, 'teams win, individuals lose.'" To promote teamwork, companies must provide a team goal and make sure that every individual meets his or her needed sales results; otherwise, the team goal may not be met.

12. Lacking Support and Customer Care That is Second to None

    In today's highly demanding consumer environment, sales support and Customer Care™ are just as important as selling a great product or service. We all know people who have canceled a contract with a well-known manufacturer, not because of product performances, but because of lousy Customer Care™ and customer service. A situation like this qualifies as a sales prevention department because the unhappy customer is not likely to buy anything from your company if you have that kind of problem.

13. Being Rude and Other Displaying Unprofessional Behavior

    It is management's paramount responsibility to train and communicate clearly with the entire company that rudeness and unprofessional behavior, either within the company or with customers, is totally and categorically frowned upon and not tolerated by management.

Focusing on the Strategic Needs of Your Customers

In addition to avoiding the above problems, a progressive company with savvy management must focus completely on customer needs and Customer Care™. Today's customers are looking for the following:

1. Better, Cheaper, Faster

   This is 5clearly the formula for success for any progressive company. Customers demand better, cheaper, faster products. To survive, suppliers must comply at all costs.

2. Customers Need a Competitive Advantage

Obviously, with the highly competitive environment today, if you don't have the competitive advantage, you cannot sell your products. Customers expect their vendors to give them a competitive advantage and, most importantly, they expect their vendors to differentiate their products from their competition.

3. The Customer Demands All of the Above, Yesterday

   That is, they want it and they want it now. To prosper, you need to reinvent your company to comply with all of the requirements indicated above to run a successful company and avoid lost sales.

These rules were created as the result of years of experience in business. For us, these are the cardinal rules of any company. Indeed, over the years, we have personally lost a lot of money because of ignorance of the above golden rules of business, and our objective is to share them with you so you don't make the same mistakes that we have.

The bottom line is that if you want to eliminate the sales prevention department, which exists in perhaps 95 percent of companies, you need to address all these problems and eliminate anything that is contributing to sales prevention in your company.

Copywriting: The Art of Merchandising through Writing

Copywriting is the art of convincing your reader to take a specific action. And yes, it's still copywriting if it takes place in a podcast or video, if you're doing it well.

For copywriting success, refer to the following 69 Common Sense Rules of Marketing.

THE 69 COMMON SENSE RULES OF MARKETING

1. The Paramount Rule - The purpose of a promotional message is:

   - To inform

- To communicate the benefits
- To call for action and response in a professional way

2. Benefit driven – What's in it for target audience?

3. Attention grabbing instantly

4. Muhammed Ali style – confident, going against grain.

5. Be outrageous.

6. Stand out.

7. Headline should scream at buyer while naming the benefits in five words or less (see #35), i.e. BOOST PROFITS WITH ________!

8. Drive home the benefits over and over again. Forget worries of redundancy!!

9. Promotional messages and ads must be concise, well laid out.

10. Use plenty of white space.

11. Do mass customization, which is the essence of database marketing.

12. Practically all promo pieces must be able to be sent via PDF through e-mail, available on website and through social media.

13. Answer who, what, when, where, why, and how (to participate, register, get more info, etc.).

14. If offering a toll-free number, it must be at least ¼ inches tall.

15. Offer savings and value added and ask for the order.

16. Headlines must be upper case/lower case – not all caps.

17. DO NOT use Sans Serif font for body text (avoid light font faces). Choose a functional font that is legible even in the four-color version.

18. Use yellows, warm reds, orange and magenta as primary colors with blue, green, etc. as secondary colors. Never use brown, dark greens or gray; they are turn off colors.

19. Proofing is not the time or place for defining creative marketing strategy. Staff members reviewing final copy text are searching only for mistakes and omissions and should not be suggesting copy changes. The proofing process should NOT take more than 2 hours.

20. A promotional piece must sell the product – it does not have to win the Nobel Prize for literature!

21. List the facts in a benefit-driven way.

22. **In copywriting, the first 10 words are more important than the next 10,000 words.**

23. In long copy, repeat benefits several times.

24. In short copy, repeat benefits twice in different ways.

25. Use colors judiciously. Do not over use colors – use screens, pastels of yellow, green, pink or blue to accent. Never use gray!

26. Excessive color hinders reading.

27. Avoid full bleeds or solid colors on full page with drop out type. They are difficult to read.

28. The four most powerful words in copywriting are **FREE, NEW, ANNOUNCING** and **SAVE $XXX** - use them **OFTEN**.

29. The words **FREE, NEW**, and **ANNOUNCING** must be gigantic, all caps unless in headline.

30. Use fear in copy (i.e., "You must market or you won't survive.").

31. Use plays on words (i.e., "Man does not live on bread alone – he needs marketing to survive.").

32. Always give multiple ways to respond: phone, postal mail, e-mail, website address or social media.

33. Always give **FULL** postal address (include USA). Also add phone (toll-free and direct number) and website on EVERY page. Always spell out state name, do not abbreviate.

34. Avoid direct mail whenever possible; use e-mail first. Today, most people do not read direct mail.

35. Headlines must never exceed five words and must be huge (32 pt.) + super bold (and benefit driven). (i.e., "Subscribe Now and Save," or "Become a Selling Genius.")

36. Use innovation and be cost effective. (i.e., By far **the** most memorable **Christmas card we ever received cost ONLY one cent to produce**!)

37. For copyright protection, Page 2 and all subsequent pages of all promotion materials must include Copyright © [prevailing year] with full name of company (No abbreviations).

38. Always use company logo.

39. **ALWAYS USE ™ OR ® for trademarks AS APPROPRIATE** on every page.

40. In marketing, timing is everything – without exception, **ALL** promo pieces <u>**MUST**</u> be approved by marketing vice president and one other editor or similar title and **MAILED ON TIME**.

41. Missed deadlines can cost a company hundreds of thousands of dollars in revenue. Therefore, <u>**NO EXCUSES**</u> are accepted for missed deadlines.

42. Read and study **ALL** promotional materials, websites, social media postings, etc. to ensure you have a **FULL** understanding of your business, or you will not be able to write an effective promo piece.

43. Never talk down to the readers. Begin by saying: "As you know".

44. <u>**Small Type Guidelines**</u>
    a. Avoid using a type size smaller than nine or 10 points. In this case, the column width must not exceed 2 1/2 inches with at least 9/11 leting.
    b. <u>**NEVER**</u> run small type across the page.

45. <u>**PROPORTIONS:**</u> Correct type size proportions are vital to the success of the marketing piece. For example, <u>**benefits type size should be at least six to seven times larger than dates or location of an event.**</u>

46. Remember, when you emphasize <u>**everything, you are emphasizing**</u> **NOTHING!**

47. <u>**ALWAYS USE PHOTOS, CHARTS, GRAPHS, AND OTHER ATTENTION GRABBING, PLEASING VISUALS.**</u>

    If judiciously and thoughtfully selected, they will make the promo piece <u>**more inviting to read**</u> and <u>**10 times more effective**</u>.

48. <u>**REMEMBER, NOTHING TURNS OFF A READER FASTER THAN WALL-TO- WALL TEXT!!**</u>

49. Be consistent and conservative with your choice of type styles and colors! Too many type styles and colors cheapen the work and makes it less inviting to read.

50. **IMPORTANT**
    To avoid **wasted time** and to speed up the process, **BE SURE** that **ALL** promotional materials **FULLY** comply with **ALL** of the above, **BEFORE** they are routed or given to a marketing vice president.

51. Please note that NO marketing or promotional materials should be printed without pre-approval of a department head and not until all items in #66 are done. There should be **NO EXCEPTIONS**.

52. Printing the **"RIGHT QUANTITY"** of materials.
    Believe it or not as simple as this matter should be, over the course of a year companies often print large volumes of materials for absolutely no reason and due to neglect. Loss to a company can be thousands of dollars. It is the marketing personnel's paramount responsibility to work extremely closely with other departments to ensure waste is eliminated.

53. **When is a Marketing Project Completed?**
    Answer: When the material is printed or prepared to be placed online with outstanding quality and **IF AND ONLY IF IT IS LABELED WITH THE PROPER DISTRIBUTION LIST AND IT IS IN FACT DISTRIBUTED.**

54. **ALWAYS FOLLOW UP DAILY** with any design or distribution partners to make sure your project is **NOT** set aside because another customer is screaming louder!

55. **ALWAYS INSIST ON RECEIVING THE ORIGINAL COPIES OF POSTAL RECEIPTS** to make sure your **JOB WAS IN FACT MAILED ON TIME**.

56. Make sure target distribution lists are ordered at least two weeks in advance so they arrive **ON TIME** for the project.

57. **Know the IMPORTANCE OF FOLLOW THROUGH.** Persistent daily (or even more frequently) follow through at **every step** is **THE KEY** to meeting **MARKETING DEADLINES**.

58. To be a leading Marketing organization, your marketing materials **MUST** be the **BEST** and **MOST** attention grabbing on every executive's desk, computer screen and/or mobile device.

59. **AVOID** design and layout and excessive copy that would make a marketing piece too busy. **NOBODY READS** anything that is too busy.

60. **THE BEST ADS** we have ever seen had the following in common:
    a. A great headline that would literally force you to read on.
    b. Outrageous and extraordinary or enormously simple visuals. i.e., the Egg ad.
    c. A message that was extraordinary and memorable. i.e., "IF YOU THINK STAYING IN SCHOOL IS BORING, TRY THE UNEMPLOYMENT LINE." Graphics showed a classroom and an unemployment line!! AND NOTHING ELSE!!

61. **NEVER - NEVER** use color to print small type sentences or paragraphs – **BLACK** is the only color that is most legible in small type and always such type must be against a white background. **Exception**- single words such as: **FREE and NEW** must be in color (in red and always bold and big (B + B).

62. All press runs must always be approved by a marketing vice president and the division head via a purchase order, and must be accompanied with agreed upon distribution lists.

63. Be sure that **ALL PERTINENT** corrections are **IN FACT** carried out.

64. **<u>DISTRIBUTION LIST ACCURACY</u>** – when renting lists, be sure that you have <u>in fact received the titles you have ordered</u>; otherwise, you are wasting your time and your company's money.

65. **<u>EXPECTED COMPLETION DATES for print projects</u>**
With excellent planning and coordination with other departments, the following should be used for EXPECTED COMPLETION DATES:

    a. One-Page Flyer:
        - copy – ½ day
        - layout (B+W)
        - approval – two hours
        - to printer – next day (end of day)

    b. Four-Page Brochure: in <u>three</u> days to printer.

    c. Eight-Page Brochure (Conference Brochure): <u>one week</u> maximum.

    d. Larger size brochures as specified by creative team and a marketing vice president.

66. **<u>COST CONTROL – YOUR RESPONSIBILITY</u>**

    a. Use most economic size vs. paper sizes available.
    b. Most economic mailing size – get information from two local Post Offices for accuracy.
    c. Assist the creative team for selecting the Best Quality printer, the Best Location, the Best Price, and Most Importantly, reliability and quality of printing.

The above must be done before a printer is chosen. Also, for every job, three price quotes from printers will be necessary. Also, a purchase order must be signed by the authorized person before the printer is advised of the press run.

66. **Important Final Step**

    It is the marketing department's responsibility to communicate without exception, every single advertisement and promotional piece prepared with all pertinent members of the company, particularly sales, marketing, call center, and all department heads.

    In addition, it is marketing's responsibility to work with the creative team and immediately convert to PDF all promo pieces, brochures, media kit materials, etc., to be e-mailed to appropriate customer contact lists, added to website and promoted through social media.

67. **The Real Objective**

    The real objective is to produce the most effective marketing and promotional messaging, in a time sensitive manner, with the best quality at the lowest cost.

68. **IT IS THE MARKETING DEPARTMENT'S RESPONSIBILTY TO MAKE SURE THAT ALL OF THE ABOVE GUIDELINES ARE 100 Percent CARRIED OUT**. YOU MUST BE PROACTIVE – GET INVOLVED – LEARN ALL YOU CAN ABOUT EVERY DEPARTMENT of the company, the industries served, and about integrated marketing. Your success depends on your work ethic, and your FLEXIBLE, POSITIVE, CAN DO ATTITUDE!

Chapter 7

Loyalty Marketing – Customer Retention Depends Upon Customer Care™

Loyalty Marketing....
Because Companies Live or Die by Repeat Business

Historically, when upgrading customer service first appeared as a business strategy, it was perceived as something you might be obliged to do if your quality was inferior. Remember, customer service became a business priority in the '80s, when American products were being overshadowed by high-quality counterparts from Japan and elsewhere. Yet, now that American products are seen as having caught up in terms of quality, customer service is still a hot topic.

In fact, approaches to customer service have been elaborated to form Customer Care™ and customer retention strategies, which in turn are leading to relationship marketing and loyalty marketing. All these strategies have one thing in common: **They all recognize it is far more expensive to gain a new customer than it is to keep an existing one.**

But what about quality? If you have top quality, do you still need to pursue relationship and loyalty marketing schemes?

In a word, yes! In the emerging world economy, where high quality is more and more taken for granted, customer retention and customer loyalty promise to be important competitive differentiators. **Perhaps the best way to keep the importance of customer loyalty in perspective is to remember that 75 percent of the purchasing decision is based on emotion.** Now, few emotions are as potent as those aroused by loyalty. If you can cultivate customer loyalty, and combine that with high quality, you will have an unbeatable combination in today's super-competitive business environment.

What Is Loyalty Marketing?

There are at least two general approaches to loyalty marketing:

I. **Focus on Your Customer's Customer.** This method is particularly appropriate for a service provider. To be successful at this, you'll need to understand your customer's business exceptionally well so you can recommend methods and innovative ideas by which your customer can impart loyalty to its customers. If you can do this, you will keep your customer for life.

II. **Focus 100 Percent on Your Customer.** There are many things you can do to impart loyalty to your customer. Among them are the following:

    1. **Relationship building:** Given 75 percent of the buying decision is based on emotion, there is no substitute for relationship building.

    2. **Systematic individualized Customer Care™:** You can let a customer know you care by being aware of his or her interests and paying attention to important events in the customer's life.

       You can, for example:

       - Send a birthday card to every customer.
       - Send a thank-you note after every major order.
       - Send flowers on happy and on sad occasions.
       - Send appropriate gifts. (For example, if your customer plays golf, you could give him or her a set of golf balls or a free subscription to a golf magazine.)

We could go on and on. Let your imagination take over, and you will figure out similar appreciation techniques.

Rewards and Referrals

Customers who will refer you to others are extremely valuable. Show them extra appreciation and reward them for their repeat loyalty. For example, consider a rug-cleaning operation.

After a rug-cleaning job is completed, the company could follow-up a week later to ask if the service was performed satisfactorily. If the customer is unhappy, offer to do whatever is necessary to satisfy the customer – free of charge. If the customer is unhappy, offer to do whatever is necessary to satisfy the customer – free of charge. If the customer is pleased, you could ask him or her to recommend your service to others.

For every referral, the customer would get 25 percent of the next cleaning job. And, after every four referrals, the customer would get a certificate for a free rug shampoo to be used anytime the customer wanted it.

Of course, implementing such a program could increase overhead, but the investment would be well worth it. After all, it costs 10 times less to keep an existing customer than it does to gain a new one.

The Inevitability of Loyalty Marketing

In the future, loyalty marketing will no longer be the exception, but the rule. Why? Because companies that ignore loyalty marketing will go out of business, to be replaced by competitors who know how to take care of their customers.

Here is a simple example of how quickly customer loyalty can be undermined by careless customer service.

While attempting to buy coffee at a fast food restaurant, Nadji had a dispute over the amount of change he should receive. Evidently, the cashier had made the mistake of thinking the 20-dollar bill he had given her was only a five.

What happened next? Well, the restaurant's employees handled the situation about as badly as you could imagine. He was kept waiting for about half an hour while they counted all the money in the cash register. When they finally found that the register did, in fact, contain 15 dollars more than it should have, they begrudgingly gave him the change.

Yet despite having kept Nadji waiting because of their mistake, and despite having subjected him to surly behavior, no one saw fit to offer an apology. And no one thought to give him fresh coffee, the first one having long gone cold.

Clearly, this is the worst possible outcome for the customer experience, and – needless to say – he will never patronize that restaurant again.

Companies Cannot Afford to Let Employees Abuse Customers.

Mistreated customers will not waste their time trying to teach your employees the rudiments of Customer Care™. On the contrary, they will simply, vote with their feet and take their business elsewhere.

We chose to vote with our feet with a major car rental company. This company has forever lost our company business (which is considerable) because of the indignities we have endured at the hands of that company's employees.

The Importance of Employee Satisfaction

The anecdotes just cited indicate that surly employees can destroy your business. If you need more convincing, consider this statistic from a recent Nielsen survey: 68 percent of customers who stop doing business with you change vendors because of indifference or hostility expressed by an employee of your company. **Just think: Employees with bad attitudes account for more lost business than all other reasons combined!**

So what can we do to enhance employee satisfaction? Continue investing in quality.

As noted by W. Edwards Deming, the celebrated management expert, most employees want to do high-quality work. Indeed, few things are more frustrating to a good worker than being prevented from doing a good job. To ensure employee satisfaction, which leads to customer satisfaction, you must give your employees the training and tools they need to get the job done right.

And, while you are at it, make sure you pay attention to vendor satisfaction. Poor performance by a vendor can make you look like a second-class operator through no fault of your own.

The Use of The Phone

The telephone remains an important and powerful tool to instill customer loyalty, especially for your top-tier customers.

- Call customers periodically to thank them for their business and gauge satisfaction.
- Call customers to wish them happy birthday or ask about their interests (sports, for example).
- Call customers to give them information or items of particular importance to them. For example, for a customer expressing interest in sports, you could call and offer tickets to a game or offer to send a t-shirt or other apparel item with the logo of their favorite team.

Use your imagination. The idea is to use the power of the phone for communication on topics that will enhance your relationship. You will surprise customers with a call of loyalty, but it will not be forgotten!

A Few Well-Known Loyalty Marketing Techniques

Examples of how to earn customer loyalty abound. Here are a few well-known examples:

- Airlines grants frequent flyer miles to customers.
- Companies sell breakfast cereals that include coupons in the cereal boxes.
- Car makers offer car buyers access to free roadside service via 800 numbers.

This last example which is more service oriented, can be elaborated further. For example, a major motorist club, such as the AAA, could approach a car maker and offer its services at a discount. This way, the car maker could offer its customers free road service for, say, the first 50,000 miles of car ownership.

Such an arrangement could offer car buyers great peace of mind. It would thus enhance both the buying and the loyalty processes. Everyone would win: The AAA would make some money by selling its services to the car maker; the car maker would gain some customer loyalty; and the customer would get some peace of mind.

We encourage you to think of similar scenarios that apply to your business. If, in your business, you develop offerings such as those described above, and make them a part of every transaction, you can count on customer loyalty. You will have created a bond of loyalty. This is an outstanding achievement. After all, loyalty takes us into the realm of emotions. And emotions, as we have noted, account for 75 percent of the purchasing decisions.

The challenge is for vendors to think of ways for clients to make the client's customers happy. Of course, there will always be a need to acquire new business, and adequate Customer Care™ and customer service will always be important, but it is loyalty marketing that will separate the winners from the losers in today's marketplace.

Why? Because companies live or die by repeat business.

8 Steps For: Taking Customer Loyalty to the Next Level Because Companies Live or Die from Repeat Business

The Vital Necessity of Customer Loyalty

Conventional wisdom dictates that customer loyalty is the most vital aspect of every business because companies live or die from repeat business. No company can exist without customer loyalty and retention.

The following eight guidelines are based on the principles that will enhance your customer loyalty. Consider how you may incorporate each one of them into your organization.

1. **Keep In Touch Frequently.**
   It has been proven that if you do not keep in touch with any account for more than 30 days, you do not own that account. Your salespeople must live and breathe with this principle.

2. **Build the Relationship.**
   Unfortunately, many salespeople just want to get the business now and worry about the relationship later. That kind of philosophy is doomed to fail because customers are smart and can figure out whether or not you are sincere about your relationship with them. We can state emphatically that business is all about relationships. If you don't have them, you have no business.

3. **Nurture the Relationship.**
   Much goes into nurturing the relationship once you have established a relationship. Unlike the common belief, the relationship will not be nurtured if you constantly ignore the customer's needs and interests. The only way to nurture them is to show genuine interest that you're concerned about addressing their interests and needs on a regular and continuous basis. And, there is no shortcut in this concept.

4. **Manage Your Customers' Expectations.**
   One of the most common mistakes made by salespeople is to over-promise and under-deliver. This phenomenon must be completely and categorically eliminated in every sales department.

   You need to make it abundantly clear that without exception, each and every member of the sales, marketing or any other department, for that matter, must live with the above principles; otherwise, customer loyalty and retention would be nothing more than wishful thinking.

5. **Anticipate Your Customers' Needs and Do Something About Them.**
   One does not keep a customer by simply forgetting about him or her once the transaction has taken place. A smart salesperson will anticipate the needs of his or her customers on an ongoing basis and, more importantly, do something about it. There are too many salespeople out there who promise everything and deliver nothing, let alone anticipate the needs and respond to those needs.

   We are sure that practically every sales department has a few people like that, all of whom should be terminated, as they have no value and they can only serve as a cancer to the sales department.

6. **Hire Sales Staff With Character.**
   In one of the recent documentaries about successful coaches in pro football, Joe Gibbs, the famous coach of the Washington Redskins, was asked the question, "What criteria do you value the highest before you select a football player for your team?" Joe Gibbs responded, "Character is by far the most important attribute that I look for when selecting team members for my team." A reporter asked him, "How do you define character?" He stated, "I want people who care the most about what is in the best interest of the team and not about what is in the football player's best interest."

Therein lies a tremendous amount of wisdom that every employee of every company must respect. Salespeople who care only about their own pockets have no place in any corporation in today's highly competitive and customer-savvy environment. If the salesperson does not care about his team or his company, the customers or prospects will have no reason to do business with that salesperson. Consequently, there is no reason for that salesperson to exist in any company. As simple as this sounds, one always encounters salespeople who don't give a damn about what is in the best interest of their companies; they care only about their own selfish gains.

7. **Go the Extra Mile.**
   In today's globally competitive environment, one of the most powerful attributes that can separate your company from others is if your sales and customer service staff to go the extra mile. Customers will always remember if your sales or customer service department really cares about them or their business. If the representatives of your company do not show a genuine interest in helping the customer on a continuous basis, then there is absolutely no foundation for customer loyalty.

8. **Help Your Customers Save Money.**
   By showing the customer how to reduce costs by recommending ideas and new processes or know-how, the customer might change or adopt to save considerable cost. This type of concern goes a long way toward generating loyalty.

The Anatomy of Customer Acquisition to Customer Loyalty

Given that customer loyalty and retention are the life-blood of every corporation, it behooves us to look at the anatomy of customer acquisition to get a greater understanding of the process.

Marketing → Awareness → Lead Generation →
Sales → Customers → Profits →
and continued sales and profits (reoccurring revenues) comes from: →
Customer Loyalty

The diagram above details the process that initiates customer acquisition and follows through to customer loyalty.

In Business, Re-occurring Revenues Are The Only Thing!

Vince Lombardi, the legendary coach of the Green Bay Packers pro football team, whose team won several Super Bowl championships, once said, "Winning isn't everything, it's the only thing." This great principle of business, coming from one of the greatest, if not the greatest, football coach of all time, is equally applicable not only to winning companies, but also to all companies that wish to survive in this highly competitive economy.

Effective Advertising Begins with Innovation.

In addition to reaching the right people, any advertising program must be innovative if it is going to make an impact on your audience. It has been said that an average person is exposed to in excess of 1,000 advertisements per month. Consequently, the only thing that will stay in anyone's mind is the ad that is truly innovative.

We think every person of reason would agree that effective advertising is vital to the survival and growth of every organization. In business, there are numerous success stories where advertisements have made companies successful, and a lack of it has prevented growth or even the existence of some companies.

We need to apply the same principle to all other advertisements. In other words, if you are unique and innovative, you will get attention and new business from your ads. If, however, your advertisement is the usual boring and ineffective style, nothing will come out of it. The sad fact in this case is that those who place useless ads and get no response don't blame the ad, they always blame the concept of advertising by saying, "Advertising doesn't work for us!"

Integrated Marketing Is The Only Way To Go.

In today's highly sophisticated advertising environment, the only effective way to market is to use the concept of integrated marketing, which incorporates print, e-mail, digital, social, trade shows and the assorted other components of integrated marketing.

If you understand that, you cannot lose sight of the fact that your advertisements must be innovative and benefit-driven and must clearly differentiate your products and services from your competition. If these guidelines are followed, then success and qualified lead generation will follow.

We felt that describing the process of lead generation, which is a direct result of effective advertising, was necessary to understand how difficult it is to get qualified sales leads (which lead to customer acquisition) to help us appreciate how hard we have to work to generate customer loyalty and customer retention.

Before We Go Out Of Business, Maybe We Should Think About CRL!

We have discussed the vitally important role of generating new sales leads from which new customers result. Between 50 to 70 percent of customers are lost for a variety of reasons, including natural attrition, the state of the economy, poor management, technological obsolescence, etc. Procuring new business is the lifeblood of any enterprise wishing to survive, especially in difficult economic conditions. Ironically, many managers do not realize that new business or new customers are vital to the survival of any business.

By cutting advertising, marketing, promotions and trade show participation, they end up cutting the source of new customers, which is equivalent to shutting off the oxygen line to a patient. Believe it or not, as foolish as the above scenario may sound, there are many companies that have stopped all expenditures and promotional activities to generate new business and new customers. Having said that, it becomes obvious that customers are exclusively the source of business survival.

Going back even to 1982, when Telemarketing® magazine in a pioneering effort laid the foundation for what is now the customer interaction, contact center and call center industries, Nadji repeatedly said, "Businesses live or die from repeat business."

Today, nothing holds true more than the above fact. In other words, here we are many years later and the customer is still the king. We are reminded of a great French philosopher who said, "The more things change, the more they stay the same"!

The Foundation for True CRM

CRM cannot exist without ERM and VRM

Without effective employee relationship management and vendor relationship management, CRM is little more than wishful thinking. That fact was true then, it is true today and will be true forever. The ironic thing is that while so many talk about customer service, customer relationship management, etc., no one seems to talk about employee relationship management or vendor relationship management.Make no mistake about it, there is no room for shortcuts or tolerance in this basic fact of business life in the new millennium.

Dot Com Doom

The early demise of a great many dot com digital companies came as a result of extremely poor customer service, poor fulfillment and total disregard for the wishes of the customer. And we all know what happened. Indeed, many of the dot com companies vanished because they chose to ignore the customer either by ignoring the input of customers or offering products no one needed or delivering products too late.

The message is loud and clear. Cherish the customer, listen to the customer, pay attention to his or her needs or vanish.

The Mass Confusion over the Definition of CRM

The greatest mission of CRM is to provide customer loyalty and retention. We have stated that the customer is precious, the customer is the king and the smartest thing any business can do is to open the channels of communication with customers and thereby provide the kind of services customers need or can use. To describe the final results or definition of

CRM, we would have to say CRM refers to any and all activities that promote Customer Care™, loyalty and retention. When such a condition exists, there will be repeat business, which provides credence to the mantra: "Businesses live or die from repeat business."

Being #1 Is Everything, Except...

In any business, it is vitally important for any innovative company with savvy management to be No. 1 in its line of business. This fact of life can be proven very simply as follows:

We've asked you before, but let's review one last time because this is so important for you to remember.

Case #1 – What was the name of the man who first flew over the Atlantic? The answer, of course, is Charles Lindbergh, but if you ask anyone who the second man was to fly over the Atlantic, no one remembers and no one cares for No. 2!

Case #2 – What was the name of the horse that won the Triple Crown in 1973 while breaking all speed records? The answer, of course, is Secretariat. If you ask the name of the horse that was always No. 2 directly behind Secretariat, no one remembers the name of that horse because no one cares for No. 2.

Case #3 – Who was the first man to step foot on the moon? The answer, of course, is Neil Armstrong. Now if you ask who was the second person who landed on the moon, many are hard-pressed to answer.

Putting it in Perspective

Obviously, it takes an exceptionally high level of creativity and innovation and business development savvy to take a concept and develop it into a No. 1 position in the industry. However, if you choose to ignore the wishes of the customer, you can fall from No. 1 to No. 2.

How About a New Department Called CRL?

In the mid '80s, awareness led many companies to provide better and better customer service. While customer service, per say, should be the only sustainable competitive advantage, today, it has evolved into customer loyalty, customer retention and overall customer satisfaction. In plain English, no company, regardless of how great its products and services may be, can survive in today's business by ignoring the principles of Customer Care™, customer retention and customer loyalty.

So why not establish a new department in your company called the Customer Relationship and Loyalty department, or CRL for short? Then assign a top executive in the company called Chief Loyalty Officer, or CLO for short. You may think this is really a far-fetched idea, but if you think about it and if you are really serious about providing outstanding CRL or customer relationship loyalty services, there is really no shortcut to success.

So before we go out of business, maybe we should think about CRL!

Chapter 8

The Do's and Don'ts of Social Marketing – From Followers to Brand Ambassadors

First Steps to Get Started:

Because social media is such a new form of communication, it is uncharted territory for many business owners, and even some seasoned marketing professionals. From where to start to what to measure, there's a lot to consider when crafting your social media plan.

So where do you start?

The first step is to consider your goals. What are you trying to accomplish through social media and how does it tie in with your overall marketing plan? By knowing your goals, it'll help you recognize how you'll be using social media for your business. Some common uses are for customer service inquiries, to share content from your blog, to give your brand more of a recognizable and personable voice, to gain share of market and to improve your SEO. Keep in mind that your goals can change over time. Once you establish them, you're not locked into this same set forever. Don't be afraid to set both short-term and long-term objectives to best suit the needs of your business.

Building your social media calendar is your next step. It'll help you stay on track and on target with your goals. I typically build out my calendar in an Excel doc, creating an outline of either specific posts that will be going out each day, or an overall idea or theme for each day. Once your posts are ready to go, you can do a bulk upload right from your document into a publishing platform like Hootsuite. Once you get into a steady rhythm with your social media pages and campaigns, you may choose to post directly to your publishing platform. Until then, a calendar is a great way to stay on track.

Best Practices

When it comes to best practices, I can go on for days. But the most basic way to break it down is to simply use common sense!

* Don't be spammy! When you're on social media, you most likely don't want your entire feed filled with overly-promotional material. Your followers feel the same way. Try to abide by the 80-20 rule: 80 percent of your content should provide value to your followers, 20 percent can be promotional.
* Don't be offensive. It sounds like a given, but you would be surprised how many posts you see that don't take this into account. Try to steer away from sensitive issues like politics, religion, and other personal beliefs that may divide an audience.
* Try to interact with individuals one-on-one when possible. If someone mentions you in a post or retweets you, thank them either through a personal message or liking the post. If someone asks you a question, respond as quickly as possible. Because of the fast-paced nature of social media, users expect their questions to be addressed almost instantaneously. We know your schedule is already packed and an instant response may not always be possible, but try to set a response time of 24 hours.
* Provide interesting content that brings value to your followers. This one is a close cousin to not spamming your audience. When crafting your social media content, try to keep in mind what your audience will get out of it. Valuable content comes in many forms. It can be through discounts, through providing up-to-date information on your industry, through fun or funny posts, or images/videos that engage your followers. Give them a reason to come back for more.
* Use photos, videos, and animated gifs when possible to take more real estate on the page and make your posts more engaging. Posts

that include a photo or video have been proven to have a higher engagement rate than those without. As the saying goes, a picture is worth a thousand words. And when you're working with a limited number of characters, that can mean a lot.

* Keep your posts concise. Twitter limits you to 140 characters or less. Even though Facebook and LinkedIn don't, that doesn't mean you should write a novel on these platforms. Users have a shorter attention span than ever, so try to keep it short and sweet.
* Pay attention to best times to tweet using tools like Hootsuite and Tweriod. These tools will look at the times that most of your followers are active and engaged and post during those times. Not all social media sites have tools like these available, but most have some stats available. Test at different times and days of the week and run your own experiments on when is the optimal time to post.
* Be consistent! This is one of the most important yet oftentimes overlooked step. Post often and regularly. Blogs and social media pages that thrive are persistent. They keep posting consistently for years, where others give up after just months.

Crafting Effective Posts

When crafting your posts, keep a few things in mind:

* Numbers and statistics typically lead to more clicks and shares/retweets, especially if you're in a B2B environment. But don't bombard your followers with a constant stream of stats – include some variety.
* The second and third most popular types of posts in terms of engagement are posts that address the reader directly and how-to posts, respectively. When you read a post, ask yourself if you would want to interact with it if you weren't the one writing it, and if it will resonate with your target audience.
* Include hashtags! It will help your content get found. Your entire post should not be made up entirely of hashtags, but include a few that make the most sense.

Measurement

Once you have a solid few weeks of posting, start to measure what's working and what's not. Look at the number of shares, retweets, likes, clicks, and comments for each post. Do the most highly engaged posts have anything in common that you can continue using in future posts? Was a different style of post a complete flop? Use these measurements to guide you moving forward and continuously monitor results. I like to review and analyze stats weekly or monthly at the very least.

A social network comparison:

Knowing what makes each platform unique will help you determine where on social media you should devote your time. I'm a big believer in not having your brand on every platform imaginable, but managing a few platforms well that work best for your business. While there are more social media outlets available now than ever before, there are a few that universally work best for business: LinkedIn, Facebook, Twitter and Instagram.

Built on the idea of professional networking, LinkedIn is a natural choice for business-to-business interactions. Build your company's brand using a company page, or specialized groups to encourage discussion and networking. It's equally important for your employees, your sales staff in particular, to be active on the platform. It's a great way to keep in touch with clients, and to search for and connect with potential new business.

Twitter is the fast-paced, 140-character network that combines businesses, professionals, consumers and everyone in between. Twitter can be an effective avenue to connect with people who you may not have access to otherwise. It's not just important for B2B in connecting with potential customers, but business to consumer as well. Customers tend to take to Twitter to either blast a company for its awful customer service, or praise a company for the great experience they had (there's rarely an in-between). It's important to be able to monitor this chatter and use it to your advantage and use negative Tweets as an opportunity to change their opinion on your brand. Whether you're reimbursing your customer for a bad experience or just letting them know that their opinion is being heard, social engagement is essential to your brand. If your customers are praising your name, acknowledge that too and make them feel the love right back. It may result in a customer for life. Since Twitter does move so quickly and rarely ever sleeps, your activity will be essential. Aside from social customer service, Twitter is also a great place to establish yourself as an industry thought leader. Share content that you've created to drive traffic to your blog, curate content from around the web that is related to your industry (be sure to give attribution when you go this route), retweet other leaders in the space, and start conversations when you see fit. If you're not tweeting at least once a day, you're not tweeting enough. As we talked about earlier, there are tools to help you gauge the best times to tweet for your specific followers and the best days to be active. Use this as a guideline to get most impressions on your posts. There are also tools available like Hootsuite and HubSpot to monitor chatter around your brand. You can set up specific keywords to populate into a feed so that you can keep a pulse on not only your company, but your competition and your industry as a whole.

Facebook appeals primarily to the B2C business, and Facebook company pages are the most popular way to promote your business through the platform. To build your following, it is recommended that you drive people to your Facebook page through other platforms. Another great way to build up your network on Facebook is through paid ads. There are several different options available, from retargeting to selecting a specific demographic to serve your ads to, to advertising only to your email database. Facebook makes reaching your target market accessible and relatively easy. Facebook is a great place for things like contests, photos and fun promotions. It is one of the best places for your company's personality to shine!

If you have a very visual brand, you need to be on Instagram. While in its beginning stages, Instagram appealed to a much younger demographic of mostly teens, it has slowly but surely been trickling up to older users, as Facebook did in its earlier days. Is posting your photo enough? Of course not. You'll also need to be sure to include hashtags that are related to your post. The more hashtags, the more likes you typically get. However, a massive list of hashtags can make you look like you're spamming and may turn off some of your users, so throwing together a list of hundreds probably isn't a great strategy. Try keep it somewhere under 20 for optimal growth.

From followers to brand ambassadors

Turning followers into brand ambassadors is no easy feat, no matter how large or small your brand is. Be prepared to devote time and energy if you're looking for significant results. Whether you're Coca-Cola responding to a tweet from a long-time loyalist or a small business handling customer service requests through social media, one thing is clear: Customers want, and expect, engagement. So how do you get your network – and beyond – to engage?

If you have a new product launch or awareness campaign, try a blogger outreach strategy. Many consumer-facing brands, as well as non-profit organizations, use this strategy to gain more buzz surrounding a significant event in their business.

A great place to start is to reach out to users who are already engaged and satisfied with your product/service. Ask them to participate in promotions in exchange for things like discounts, free products, or exclusive opportunities that you only offer this group. Oftentimes, brands will offer the opportunity for their brand ambassadors to get a behind-the-scenes look into their companies or will stage events to show their appreciation.

Once you've engaged your already loyal fans, try compiling a list of the top bloggers you'd like to reach. They don't necessarily need to be the bloggers with the largest networks, but those that most closely fit with your brand. I firmly believe in quality vs. quantity. Sometimes, a smaller following isn't necessarily a bad thing if those readers are devoted to the blogger and firmly trust that blogger's opinions. They can often times be a better and more influential advocate than a blogger with a huge following and not so many engaged readers. Don't forget the mom blogger community. There is network of millions of moms out there who write not only about parenting but a slew of other topics from fitness to fashion to cooking to community involvement and beyond. Once you have your outreach list compiled, send them a personalized message. Yes, you can create a general template of why you're contacting them, but be sure you add personalized details and an authentic message before you get into your pitch. Important tip: Read their blog before you reach out asking for a favor! There is nothing worse than receiving a message requesting a favor from someone who clearly has never read one of your posts. Many of these bloggers receive hundreds of requests just like yours, so do what you can to stand out. Include a note on a recent blog post that resonated with you and send follow-ups if you don't hear back. I recommend a follow-up one week after you send the initial note and a second follow-up a week or two after that if you still have not sent a response.

The goal of this type of outreach is to get more people talking positively about your brand online, allowing your brand to spread outside your own network. Be sure to regularly be in touch with your network of brand ambassadors. Have someone dedicated from your team to be in constant contact with them and build genuine relationships with them. You want them to know what's happening within your own company so they can help spread the word about it to their networks. Show your appreciation through free products, discounts, and exclusive events, but don't forget the human element. Get to know your ambassadors on a personal level – from what activities their kids like to where they're going on vacation. This simple human connection will go further than you may think in helping them feel connected to your brand.

Social media policies and protocols

Regardless of the size of your organization, it's a great idea to create guidelines around how your employees should be interacting with your brand online, and even have strategies to help your employees serve as brand ambassadors. You'll want to clearly define what can be said about your brand, and what should be avoided. You'll also want to encourage your employees to help spread your message to their own followers. One tool that's successful in helping you do this is GaggleAmp. It allows employees to easily share company messaging to their own networks.

It's also important for your employees to believe in your products and services. Demonstrate the value in your products, and they'll be more willing to pass that sentiment on to others. After all, who wants to recommend products or an organization that they don't wholeheartedly believe in? Have a designated leader who will manage the program and be there to answer any employee questions. Allow this leader to empower other employees with best practices on social media marketing. With your employees supporting you as a brand advocate, you'll get a great boost to your visibility, and create a community and loyalty within your own organization.

Which Brands are Nailing Social?

There are many brands that are doing a great job in social media, but some are really knocking it out of the park. Some of our favorites include Coca-Cola & Oreo.

Coca-Cola

Everyone knows Coca-Cola as one of America's most iconic brands. It is often looked at as a leader in marketing and branding, and its social media efforts are no exception.

The Coca-Cola social media team has a pulse on each and every mention that comes through, and almost always has a personalized response to each one – I only say almost because I'm sure there have been a couple that have slipped through the cracks. Its responses are friendly, witty, and make you feel good. I will never forget the day that my colleague wrote a blog post about Coca-Cola's branding. She shared it on Twitter and much to her surprise, got a reply shortly after. She was already a huge fan of the company, but this took her love for them to a whole new level. She expected the company's team to be so overwhelmed with mentions of its products that it would have no time to respond. The fact that the company did respond, and did so in a way that made her laugh, meant a lot and made her want to keep talking about the brand online.

We talked a lot about the importance of engagement in this chapter, and this is just one example of how that goes a long way. Coca-Cola also does a wonderful job in creating marketing campaigns that encourage content creation. Its recent "Share a Coke" campaign is one that's probably familiar to everyone. Bottles were printed with different names on them, intended to be shared with a friend with that name. For example, you may see a bottle of Coke with the name Alan on it and share it with someone named Alan. Customers were encouraged to not only share a Coke with someone, but share photos of it using hashtag #ShareaCoke. The result was customers posting photos to Twitter, Facebook, Instagram, etc., with their Coca-Cola bottle. When people shared their photos, they were not sharing it to promote Coke, but to share their photos with people they love. The result: lots of Coke-branded content and customer interaction!

Oreo

Oreo owns the space when it comes to timely posts. You may remember its "you can still dunk in the dark" tweet during the Super Bowl blackout a few years ago. The brand instantly put together an image during the power outage with a lone Oreo in the dark with text that read "You can still dunk in the dark."

The image was retweeted thousands of times within only an hour of posting. It was fun, bold, perfectly timed. Since then, it has continued to succeed in combining its beloved Oreo image with current events with its Daily Twist campaign. Each day, the Oreo team would look at what was trending and create imagery to go hand-in-hand with it. That brilliant if you ask me.

Part of the success of Oreo's social strategy is its emphasis on visuals. It's simple to find a stock photo or recycle content from somewhere else and include it in your posts. Oreo creates all of its own visuals for its social media pages, and does so in a clever, playful, and creative way. When you're scrolling through its feeds, you want to keep going and going to see what's next – it's that good.

One thing is clear from Oreo's social media posts: There is no brand with a better pulse on the now than America's favorite cookie.

What are You Waiting For?

Now that you have the tools you need, jump in and get started! You don't have to be a Coca-Cola or Oreo from day one. But with a few essential tips and tricks, you'll on your way there in no time.

Chapter 9

Hire for Marketing Aptitude, Not Just Experience

Hiring the Right People

Clearly it is critical to build a team with the proper philosophy for success. Identifying talent who are astute in real world marketing principles can certainly be a challenge.

Be sure to take it to the top in your search.

Who Needs a CEO Anyway?

Having risen from extremely modest beginnings, i.e., delivering Manhattan phone directories in New York City and getting attacked by dogs, and selling vacuum cleaners door-to-door, and later working in plastics factories, to CEO of TMC, Nadji has observed a great deal about what a CEO should be or should not be. Here are some of his views that were gained from firsthand experience.

What an Ideal CEO Should Be

A CEO, like anyone else in any corporation, must earn his or her pay. The CEO must be a strong leader and lead by example. In addition to being a good businessperson, an effective CEO should have most, if not all, of the following attributes.

1. **Must Be A Facilitator**
   Perhaps one of the most important functions of a CEO is to be a facilitator and help make the job of everyone else easier, not more difficult. Nadji has observed a number of ego-driven CEOs with no common sense who made life extremely difficult for their employees and colleagues, thereby damaging the corporation.

2. **Must Be A Visionary**
   In the dictionary, a visionary is described as one who is given to seeing visions – one who indulges fanciful theories. A visionary is one who can see a path of the future, which is not obvious at all to non-visionaries. Without a clear vision, a CEO will not be able to provide a direction for the rest of the organization to work toward.

3. **Must Sell The Vision**
   Once a vision has become worthwhile for the CEO to take action on, he or she must then sell the vision by providing convincing and compelling evidence and reasoning to the entire company so that everyone is on board and on the same page. Without this action, the vision will go nowhere.

4. **Must Keep Everyone Focused And Call For Action**
   In many organizations, when leadership is too creative and a CEO is unable to prioritize the many ideas and work on them one at a time with 100 percent focus, the ideas often get nowhere because no one in the organization has a clue as to what to do or what not to do. In many organizations, you will find that several ideas are floating around due to lack of focus and nothing gets done on any of these ideas. Thus, ideas without action are worthless. This is even worse than having no idea because the results are the same.

5. **Treat Everyone With Respect**
   A great leader must treat his or her employees, customers and vendors with the utmost respect. The greatest assets of any corporation, in alphabetical order, are customers, employees and vendors.

6. **Must Help All Employees Grow**
   As the CEO, you must be extremely prudent when hiring employees, always looking for people who have a positive, flexible, can-do attitude. You want to surround yourself with people who are true team workers, dedicated and resourceful. Once you have a team of competent, creative and ambitious people in your organization who are anxious to accept greater responsibility, it is the job of the CEO to help every employee grow and position each for greater responsibilities.

7. **Promote From Within**
   If a CEO truly follows the above, the next crucial ingredient for building a successful organization is to promote from within. We can state categorically that at our companies, some of the best performers are those who were promoted from within. If a CEO is truly interested in keeping his or her outstanding employees, nothing is more powerful than developing them, preparing them for greater responsibilities, then promoting them.

8. **Don't Let Egos Get Out Of Control And Destroy Everything**
   We have made a few observations about egos and how they affect business. On one hand, a person who does not have a strong ego will rarely, if ever, get anywhere. On the other hand, if someone has a powerful ego, but this ego is out of control, he or she will create more enemies than friends and this type of person will never get anywhere either. We have all heard the great proverb: You Can Catch More Flies With Honey Than Vinegar. Therefore, CEOs must keep their egos in check.

9. **Allow No Politics In The Organization**
   You may be thinking, "You have got to be kidding me. Every organization is loaded with politics." Our answer is politics are like a disease to any organization, and playing politics is a game played only by incompetents. Experienced leaders know this and are rewarded with stable organizations that are not likely to go out of business. A truly experienced manager does not allow politics in any organization. Period.

10. **Must Use No-Nonsense Management Practices**
    A successful CEO must keep things very simple and make it clear to all employees that the bottom line is results. Those who produce the greatest results will rise in the organization, and those who do not will not stay in the organization. We love nothing more than promoting our outstanding team members.

11. **Competent People Are Compensated Well**
    It has been said that you cannot hire first-class employees with second-class compensation. Top performers are smart people, and they know how much they are worth. It has been my personal experience that those who make very little money prior to joining you are not really bargains; they make little money because they have little to offer.

    Nadji recalls one year he hired a salesperson who had been making $17,000 a year in his previous job. Nadji was extremely reluctant to hire this person because of that, but the sales manager talked him into it by saying, "What do you have to lose?" He hired that person, and after three months, no sales were made! In short, you get what you pay for!

12. **Use The Management By Walking Around Technique**
    This should be considered one of the best parts of a CEO's job. By going around and getting to know everyone by their first name and learning something about their families, friends or pets goes a long way toward building a relationship between the CEO and the employees. Ironically, CEOs always talk about CRM, but very few talk about ERM (employee relationship management). Yet, without ERM, customer relationship management is nothing more than wishful thinking!

13. **A CEO Should Be A Door Opener For The Sales Department**
    One of the greatest things a CEO can do is to assist the sales and marketing departments to open doors to new customer opportunities.

Believe it or not, most people have tremendous respect for the high position of the CEO of any responsible organization and, therefore, having the CEO around in sales meetings with customers will produce wonders. To say the least, the CEO's presence convinces customers that you are seriously interested in their business because most CEOs do not travel with their sales departments.

Your customers will be extremely impressed and sometimes honored to have the CEO included in the sales team. In short, there is no greater door opener for new sales and no greater relationship builder in any organization than the CEO. Unfortunately, this extremely crucial function of the CEO is often neglected. Nothing could be more effective than having the CEO along to reinforce relationships with customers and communicating the CEO's appreciation for the business that the customer provides.

A Word of Caution – Taking the CEO along Could Be a Double-Edged Sword

Unfortunately, there are egomaniac CEOs who do nothing but harm if you take them along to meet customers. Such CEOs only serve to destroy customer relationships, and the company loses sales in the long run. Unfortunately, employees or sales staff cannot do very much about taking such individuals on sales calls.

Indeed, we have observed several CEOs who fall into the outrageous category who actually embarrass staff and customers and make horrible scenes in front of customers and as such, they create more problems than they are worth. If you have a CEO of that type, avoid taking him or her anywhere. In fact, you have to question the wisdom of the board of directors of such corporations that would keep that kind of CEO. Bottom line: If you are fortunate enough to have a CEO who is charismatic and compassionate, one who is blessed with a great deal of empathy and enthusiasm and is tactful, take him or her along.

There is no better way to open doors for greater sales and to reinforce customer relationship management. On the other hand, if your CEO is of the other variety, you have every right to say who needs a CEO anyway!

WIs Leads to WOs
WI = Worthless Individual
WO = Worthless Organization

NOTE: This book is designed to show you how to get meaningful and tangible results for your business success and is not focused on being politically correct. We will be frank and direct, so prepare for the facts as we see them in the reality of today's competitive business environment.

You may question why in a mainstream business dialogue the term worthless would ever be used to refer to a human being. We recognize that every individual has value and has the opportunity to be successful in the right context.

However, as business people, we must have a team that is fine-tuned and dedicated to hard work for our businesses to succeed. You simply cannot have a highly functioning organization and pay people who do not have the appropriate skills nor are willing to roll up their sleeves and put forth the effort to be successful.

We all have experienced people in companies who perform, as well as those who do not. Often there is a stark difference that is quite apparent, but sometimes WIs are able to hide in the shadows of a poorly managed organization. It is one of your primary responsibilities as a leader to pay attention to your people's performance and discern those who are doing well and those who are not.

It is usually quite easy to identify and praise your top performers; however, it can be much more difficult to single out those who are not performing and address the problems appropriately.

Today, with the explosion of new technologies, every day you come across dozens of acronyms, but in our research, no one had come up with a meaningful term for the worst of the non-performers. Let's face it, a certain percentage of the staff in practically every company is operating below the standard threshold. You must address individuals falling into this category in a tactful manner. Nadji coined the term, WI a few years back, which stand for worthless individual.

The term may be used in private conversation to refer to those who are clearly not performing. As we've stated, whether it is politically correct or not, we firmly believe you must address issues of non-performance head on and not allow problems to fester.

When ignored, the influence on other employees can begin to affect their good performance and ultimately leads to a lower work ethic overall in an organization. The acronym caught on like a ball of fire in some industries. Many C-level contacts began to use the term in private conversation to identify non-performers. The concept is quite useful without having to spell out exactly what it means.

In fact, one CEO of a major publicly traded company called Nadji to inquire regarding a specific individual he was interviewing. The applicant had provided Nadji as a personal reference on his resume. The CEO called Nadji and asked him directly, "Is this guy for real or is he a WI?" The CEO wanted to candidly and discretely determine if he should hire the guy or not. Companies waste significant resources when hiring the wrong people.

A sub-par performer is in a fact a WI for your business and should be removed promptly. You have no room for WIs in your organization unless you want your company to be a WO: a worthless organization.

TMC sales management used the term if a salesperson who joined the company would significantly under deliver and immediately be recognized as someone unable to perform to the company's expectations. Once the label of WI was attached to someone, that person had to go.

<u>Hygiene Hijinks</u>

At one of TMC's trade shows, there were a number of competing ACD providers exhibiting. TMC staff noticed that many of the companies had numerous attendees lining up to request information, while one exhibitor rarely had anyone stop by its booth. Nadji personally visited the company to figure out why it was not being successful and to provide helpful insight. As Nadji approached the salesperson, he was overcome by a most undesirable situation, a very bad odor

HIRING AND KEEPING WINNERS!

Hiring and keeping the right people is so vital, yet so difficult!

Please continue reading only if you believe that your staff is your GREATEST ASSET!

Most companies wrongly spend 98 percent of their time talking (yes, talking) about customer service, Customer Care™ and customer retention, yet are far less concerned with employee service, employee care or employee retention!

This misguided philosophy leads to excessive staff turnover and eventually will result in customer turnover, which means substantially higher costs for training new staff, not to mention loss of revenue! Once a corporation learns that without a competent staff armed with a flexible, positive, can-do attitude, there will be:

- No customer service
- No Customer Care™
- No customer retention
- No progress

Then the company is prepared to focus 100 percent on that elusive, highly complex (if not impossible) task of hiring the right people.

What Are The Characteristics Of the Right People?

The right people are those people who genuinely subscribe to the following guidelines:

- HAVE A FLEXIBLE, POSITIVE, CAN-DO ATTITUDE

from the salesperson, who apparently had never heard of deodorant. It should go without saying, but make sure your team is focused on the importance of proper personal hygiene. Customers will most certainly avoid any employee with poor hygiene. Remind your people to clean up their act.

***

- GENUINELY TRY NOT TO MAKE MISTAKES
- DON'T MAKE EXCUSES FOR MISTAKES
- DON'T MISS DEADLINES
- DO A GREAT JOB EVERY TIME
- REMEMBER EVERY DETAIL
- TAKE PRIDE IN WHAT THEY DO
- DON'T SAY, "IT'S NOT MY JOB"
- OFFER UNSURPASSED CUSTOMER CARE™
- SAY "LET ME DO IT FOR YOU"
- ARE PART OF THE SOLUTION, NOT THE PROBLEM
- ARE GREAT TEAM WORKERS
- ARE ON THE JOB EVERY DAY AND ON TIME
- SET A GREAT EXAMPLE
- DO NOT PLAY POLITICS
- ARE NOT HIGH-MAINTENANCE PEOPLE
- KNOW THAT WINNERS DO WHAT LOSERS DON'T WANT TO DO

Just How Do You Avoid Those Worthless People With Bad Attitudes?

Today's job seekers are sophisticated. Many resumes are inflated, and many candidates are well schooled in giving the right answers to your questions!

However, to discourage the wrong people, i.e., people with bad attitudes, otherwise known as WIs, we have prepared a list of statements and items designed to scare off WIs.

You may use this as a qualification test for new candidates before the interview process.

Don't Take a Job with Us If You have ever stated any of the following or any of the following actually bothers you or describes your mindset or behavior!

- You do not have any of the 44 Characteristics of the Right People (listed in next section)
- You are a high-maintenance person

- You have ever kept management in the dark about anything whatsoever
- You are a political animal. Corporate politics defined: Politics is a game played by incompetents!
- Your ego is out of control! (We don't run this company based on ego!)
- You are a clock-watcher
- You are a chronic complainer
- You are always negative
- You can't take criticism

Or if you have ever stated:

- It is not my job
- I wasn't hired to do this
- It is not my fault
- It is always someone else's fault
- I don't know (even though it is my job to know)
- I called him, there was no answer; therefore, the project died. I have never heard of being resourceful!
- I am doing someone else's job and I shouldn't be.

  Be clear with candidates that they don't want to take a job here if that are not prepared to work on 12 cylinders ALL THE TIME!

Above All, Look For Multifaceted People!

Successful companies are those that hire smart, flexible, multifaceted people. As your company grows, your requirements will change. Flexible, multifaceted people who possess a positive, can-do attitude will be able to adapt to the changes. On the other hand, negative, inflexible people will lead your company to extinction! That is why they are justifiably called WIs.

The Road To Success

The following statements represent a few of the most distinctive comments we have read about successful companies and people. They reinforce practically everything we have tried to explain in this chapter.

"The success of every organization depends on how well the entire team works together!"

"Treat people like you want to be treated, that's the key to success!"

"Your success is the direct result of your attention to detail!"

"Learn from the mistakes of others. You can never live long enough to make them all yourself!"

"Success comes from listening. I've never learned anything by talking."

Lou Holtz, head football coach at Notre Dame

Calvin Coolidge's Advice on Persistence, Determination and Hard Work

"Nothing in the world can take the place of persistence. Talent will not. Nothing is more common than unsuccessful men with talent. Genius will not. Unrewarded genius is almost a proverb. Education will not. The world is full of educated derelicts. Persistence, determination and hard work make the difference."
President Calvin Coolidge

Management Lessons Learned from Jimmy Johnson

Jimmy Johnson is one of the most successful coaches in football. "Everyone I deal with is put on a scale," Johnson said. "The better performer, the harder he works, the more he meets the guidelines, the higher he is on the scale."

"If they fall short in some of those areas, the lower they are on the scale. And they have very little margin for error near the bottom. I like guys that work hard and play good."

"But everyone in the organization has to think about team goals. Anyone who distracts from those team goals is out of line. And everyone has to understand that."

Something For Nothing

People with poor attitudes are likely to stand in front of the fireplace and say, "Give me heat then I'll add wood!" Obviously, such people will never get anywhere! Successful people don't look to get something for nothing, and they don't make money – THEY EARN IT!

Keeping the Right People

If you are blessed enough to have a great, hard-working team with a flexible, positive, can-do attitude, then your No. 1 priority should be to cherish them, treat them like gold and give them plenty of challenge, great pay and promote them within the organization.

Above all, let them know their hard work, achievements and dedication are greatly and genuinely appreciated. Let's not forget that they are indeed your greatest asset. If they are treated as such, then they will treat your customers and the company as if they are their greatest assets! After all, loyalty, respect, appreciation and the resulting success are two-way streets.

Effective Hiring and Keeping the Right People
44 Characteristics of the Right People

Successful managers are those who surround themselves with **the right people**, give them direction and then get out of the way.

The vitally important question to ask is, "Just how do you define **the right people?**

In the accompanying list, we have tried to spell out some characteristics of **the right people**.

Successful company leaders are prepared to focus 100 percent on that elusive, highly complex (if not impossible) task of hiring the right people and keeping them.

In an issue of CUSTOMER™ magazine, Ted Nardin, senior director of performance management of ClientLogic, penned an excellent article on hiring and training.

In his article, Nardin focused on effective hiring for the call center. Here are a few excerpts from his article:

"Transforming a call center from a cost-cutting afterthought to a viable brand-carrying profit-generating division of business begins by establishing effective hiring process and training programs"

"If staffed and managed properly, the contact center can be an invaluable asset to the business."

The 44 Characteristics of the Right People

The right people are those who genuinely subscribe to the following guidelines:

1. Have a flexible, positive can-do attitude
2. Genuinely try not to make mistakes
3. Don't make excuses for mistakes
4. Don't miss deadlines
5. Do a great job every time
6. Remember every detail
7. Take pride in what they do
8. Don't say, "It's not my job"
9. Say, "Let me do it for you"
10. Offer unsurpassed customer service
11. Are part of the solution, not the problem
12. Are great team workers
13. Are on the job every day on time!
14. Set a great example
15. Avoid being politicians and don't play political games
16. Are not high-maintenance people
17. Know that winners do what losers don't want to do!
18. Follow the vitally important rule of transparency
19. Never keep management in the dark about any issue, good or bad!
20. Have a GREAT work ethic
21. Believe that "the harder I work, the luckier I get"
22. Are not clock watchers
23. Come to work early, do a good job and leave late
24. While at work, they spend 95 percent of their time working exclusively on company-related projects
25. Know that if they have a problem with company policy to discuss it with management
26. Don't badmouth their employer online or on social media! That's like biting the hand that's feeding them!

27. Go out of their way to be extremely helpful to customers, other employees and team members
28. Know that if they don't like something, they should develop a BETTER solution to the problem, then inform management and persuade them that doing it the employee's way is a better way. They know that complaints do not achieve anything except lowering morale! (In which case, management is justified to terminate such people after adequate warnings.)
29. Are model employees, and are encouraged to come up with new cost-saving and/or profit-generating ideas
30. Avoid chronic complaining
31. Avoid being "bad apples"
32. Are honest and never lie
33. Never abuse expense accounts. The right people know that such abuse is equivalent to stealing.
34. Never get involved in sexual harassment
35. Have a great sense of humor
36. Are never pretenders (If they don't know the correct answer to the question, they say, "I don't know" and they don't make up answers.)
37. Don't create problems
38. Try to prevent problems
39. Spend company money like it was their own; i.e., don't waste it
40. Don't cover up mistakes and lie (Studies show that 20 percent of employees lie to management to cover up mistakes.)
41. Don't abuse company email and social media for frequent personal use
42. Are sure to under-promise and over-deliver
43. Treat customers with the highest levels of integrity, dignity, truthfulness and professionalism
44. Last but not least, they uphold company values to the fullest extent

About Attitude

The longer we live, the more we realize the impact of attitude on life. Attitude, to us, is more important than education, money, circumstances, failures, success and what other people think, say or do. It's more important than appearance, giftedness or skill. It will make or break a company, a church, or a home.

The remarkable thing is we have a choice every day regarding the attitude we embrace that day. We cannot change our past; we cannot change the fact that people act in a certain way. We cannot change the inevitable. The only thing we can do is play on the one string we have, and that is our attitude.

We are convinced that life is 10 percent what happens to me and 90 percent how I react to it. And so it is with you. The good news is that everybody is charge of their own ATTITUDE.

The Death of the Sales Manager

To do justice to this subject matter, one must make reference to Arthur Miller's legendary play entitled Death of a Salesman.

To refresh your memory, Miller won a Pulitzer Prize for his work, which he described as the tragedy of a man who gave his life, or sold it, in pursuit of the American dream. The main character in the play is Willy Loman, who, after many years on the road as a traveling salesman realizes he has been a failure as a father and husband. His sons Happy and Biff are not successful on his terms (being well-liked). Willy's main claim to fame was to use a smile and a shoeshine as the only sales technique one needs to be successful in sales.

To say the least, Arthur Miller taught us that selling success need not be at the expense of being a failure as a father or a husband. In other words, being a great salesperson and being a great father/mother and a husband/wife does not have to be mutually exclusive. For this alone, he richly deserved the Pulitzer Prize, for today, far too many people are placing business success ahead of the family and being a good father/mother and a good husband/wife.

Today we have learned that Willy's legendary smile and a shoeshine selling technique are only 5 to 10 percent of the skills needed for selling success.

Not long ago, Nadji met with the vice president of sales of one of the largest magazine printing companies, who was soliciting TMC's business. During the social conversation, he asked Nadji, "Do you promote from within?" "In fact," Nadji stated, "nearly 70 percent of the employees at TMC were promoted." He was very impressed and stated, "Then I am sure you will enjoy the following story.

"A farmer who enjoyed duck hunting had a dog that was exceptionally effective in hunting down the ducks that were shot down. The farmer, looking at the dog's performance and personality, named the dog Salesman. Pretty soon Salesman's reputation spread through the small town where the farmer lived and the town leaders called the farmer and stated, 'We understand that you have a great hunting dog and we would like to go out duck hunting with you and your dog.' And so they did.

"Of course, Salesman impressed all of the town leaders that indeed he was extremely hard working, very talented and very motivated in performing his duties. At the end of the day, everyone was grateful to the farmer.

"A week went by and the town leaders called the farmer again and asked to go out duck hunting because the last time was a fantastic event and the town leaders insisted that Salesman must come along. The farmer said, 'I would be happy to go hunting with you, except that I am sorry to tell you I shot Salesman.' People asked why. The farmer stated, 'Well, the salesman did such a good job, I promoted him to sales manager and thereafter, all he did was sit on his ass and bark all day and nothing got done!'"

There is a strong message in this story, which is not far from the truth in most cases. One of the most commonly made mistakes by management is to take a top producing salesperson and promote him or her to sales manager.The fact is, even if the salesperson was outstanding in selling, that does not mean that he or she would make a good manager.

Chapter 10

Introducing The Test for Marketing Ability™, our proprietary solution published for the first time ever!

The Test for Marketing Ability™ is presented for your use in identifying expert marketing talent for your organization. After thoroughly reading and reviewing this book, an individual should be prepared to pass the test.

The topics covered in the pages of this book will prepare the reader not only to do well in answering the test questions, but more importantly to perform successfully as a marketer for your organization.

Should you be interested or require additional support surrounding the application of the concepts presented, the authors are available for individual consultation.

THE TEST FOR MARKETING ABILITY™

Please fill out ALL blank spaces.
This is a 50-minute test.

Date____________________________Name________________________

Starting time: _____________

Finishing time: _____________

Please complete the following:

Name: ________________________________

Phone: ________________________________

Email: ________________________________

<u>PLEASE</u> PRINT LEGIBLY

Define Marketing:

Define Sales:

What are the components that make an ad effective and memorable?

1.

2.

3.

4.

5.

6.

What specific things make a flyer or a marketing brochure stand out?

1.

2.

3.

4.

5.

6.

Describe the **BEST** ad (print, radio, TV or online) you ever saw.

- When did you see it?
- Why did it stand out?

What colors turn off readers and must be avoided in designing promotional brochures or advertisements?

What colors make an ad inviting to read?

1.

2.

3.

4.

5.

6.

What color(s) must never be used **<u>for copy</u>** and why?

What information do you need to write an effective ad for a product?

1.

2.

3.

4.

5.

What is the most important aspect of an ad?

All ads must begin with a powerful ________________.

What prompts a decision maker to read an ad?

In 200 words or less, write **the most effective ad about yourself** and tell us why we should hire you as opposed to the other 20+ candidates. Begin with an attention-grabbing headline (**NOT TO EXCEED FIVE WORDS).** Use the reverse side if necessary.

Headlines

- What are they?

- How important are they?

- What do they do?

- How large or small should they be?

- What words should they contain?

- Maximum size of a headline?

Write a similar ad as you did above about yourself, except this time explain why someone should be a customer of our company. (Use the reverse side if necessary.)

In copywriting, the first ________________ are more important than the next ________________.

Define a niche market

Describe the pros and cons of a niche market.

- Pros

- Cons

GENERAL MARKETING INFORMATION

(Please answer the following questions as briefly as possible.)

What is the **first** law of positioning?

What is the **second** law of positioning?

Define positioning and explain the importance of positioning in any given marketing program.

Define database marketing.

Define database marketing in **two** words.

Define differentiation.

What is the role of differentiation in marketing?

What is regression analysis, and how is it used in marketing?

What is integrated marketing?

What is event marketing?

Define relationship marketing.

Relationship marketing is important because ____________________ percent of buying decisions are based on ________________________.

What are taglines?

Why are taglines important?

List several examples of memorable taglines you have seen:

1.

2.

3.

4.

5.

DIGITAL MARKETING

What is search engine marketing?

What is SEO?

What is the difference between SEO and search engine marketing?

Define online marketing and provide a few examples:

What is Google, and what is the significance of a Google ranking?

In a given category, which page of Google ranking is the most desirable and why?

What numerical ranking is **VITALLY IMPORTANT** on Google and why?

What is Alexa.com?

What does an Alexa.com ranking show?

Is a _______higher or _______lower number in Alexa ranking more

desirable and why? (Please check one.)

What is a blog, and what role does it play in marketing?

What is social marketing?

What is the significance of social marketing?

What is mobile marketing?

What is the significance of mobile marketing?

Good luck and best wishes

Chapter 11

Putting It All Together for Your Success

The Forgotten Art of Marketing

Even the best product in the world will not sell itself. Nadji's company, TMC, has produced the most successful trade shows in the telecommunications industry for many years. We recognize that people go to trade shows for different reasons. Some go to buy, others go to learn what the competition is doing, and others go to attend seminars and advance their expertise, while others go to learn about a new industry in which they may or may not want to enter. We even know of some who attend trade shows because of the pageantry, the glamour, the hospitality suites and the dinners and entertainment.

We have attended trade shows primarily to get involved and communicate with people at all levels of the industry, to ask them far reaching questions, to discuss trends and to make observations. We have noted a consistent problem that seems to plague all types of business industries – when all else fails, try marketing!

Insightful companies conduct market research to find a need in the marketplace and develop extensive marketing programs to sell their products long before they are ready for the marketplace. These are the companies that will ultimately be successful.

In one instance during a trade show, Nadji had a conversation with the marketing vice president of a major digital marketing services company. Upon answering an inquiry about the status of his business, he complained that it could be better. His explanation was that the company was losing customers all the time, therefore was having cash flow problems. For these reasons, he added that marketing or advertising could not be justified.

It is a known fact that through the process of attrition, which is inherent in even the best of businesses, companies lose on average 15 to 20 percent of their customer base annually. If these customers are not replaced by new customers through aggressive advertising and marketing plans, cash flow problems and loss of market share will result – just as it happened to this digital marketing services company.

Of course, businesses should make every effort to keep their existing customers engaged. Customer retention depends upon Customer Care™. Please refer to the book by the same authors, Taking Your Customer Care to the Next Level™.

Having spoken to high ranking executives from highly respected companies, it was incomprehensible to Nadji that people can actually work for such outstanding organizations while holding important positions and be so naïve when it comes to the importance of marketing for the well-being of their companies. No matter how you look at it, companies live or die not from how great their products are, but from how effective their marketing is.

Every business in every industry could also benefit from improved marketing. We are living in an information society. Let's live up to this reputation. Provide your audience (the end user) with the information it needs to make an informed decision about your company and its products. In other words, let's give marketing, the forgotten art, a chance to do what it's supposed to do: market your products and services!

Marketing By the Seat of the Pants or With Information?

As marketing becomes more and more complex, only the sophisticated marketers will survive! Those who will enjoy market leadership in the future are the managers who implement state-of-the-art marketing techniques in their companies today.

Most companies' marketing departments don't get the attention they need. Making matters worse, colleges and universities, with few exceptions, are not very eager to teach state-of-the-art marketing techniques in the classroom.

In addition, the majority of marketing consulting firms are not using sophisticated data analytics technology to generate proper information for management to make judicious marketing decisions. As a result, what you have is pure chaos – or seat-of-the-pants marketing.

Many companies that have used direct mail for years remain satisfied with one percent or less response. In fact, using rented mailing lists will rarely generate even one percent response. It is therefore critical for companies to incorporate other means to generate response. In a recent study using segmentation and customer needs as the selection criterion for a marketing campaign, TMC was able to generate 100 percent more results while reducing direct mail quantity by 50 percent!

Your marketing effort is effective when and only when it reaches the right people; otherwise, you're wasting your money no matter how large your audience. And you reach the right people by marketing based on information, not by the seat of the pants.

Reaching the right people at the right time is accomplished based on data analytics, utilizing information technology to improve the productivity and effectiveness of the marketing process. In short, data-enhanced marketing is the art of marketing with information.

A Winning Marketing Strategy Must Be Omnichannel

The phone has always been a vital part of a winning strategy to communicate directly with customers and increase market share. What was originally called telemarketing evolved with contact centers and mobile communications via online interactions.

Some of the factors that contributed to the shift from traditional direct mail to digital marketing solutions over the past decade by virtual necessity are the following.

- The extremely poor response rates received from rental mailing lists made direct mail cost prohibitive in most scenarios.
- There was a continuing rise in postal rates vs. email! To make matters worse, reportedly some 15 to 20 percent of third class mail is never delivered.
- Consumer attitudes toward unsolicited marketing messages changed.
- Government regulations mandated do not contact processes to protect consumer privacy.
- There has been growth in online interactions and mobile devices. Amazon.com and online retailers consume a constantly expanding share of total retail sales.
- Self-service solutions are preferred by a growing segment of the population.
- Social media has proliferated.

Economic survival and stiff competition necessitates that marketers continue to pursue a multichannel approach for communication with customers. Providing consistent messaging and access across a variety of channels is required today. Modern consumers are demanding that companies speak to them in the language and channel that they prefer and choose to use at any given time.

Now, more than ever before, marketers must expand their horizons and implement advanced technological data enhanced solutions to boost productivity. To survive, one must change with the times or face elimination.

The Evolution of Sales and Sales Management

We have learned that by using digital platforms, we can reach many more multiples of customers per day. Door-to-door (company-to-company) selling techniques are no longer effective without using the telephone, digital and social marketing.

From Young and Arrogant Scientist to Respectful Salesman

In his early career, Nadji was hired by one of the largest chemical companies in the world as a research chemist. This company had a great policy in which it would transfer promising young employees from department to department to give them diversified experience and prepare them for bigger and better things in the future.

Nadji once complained to his boss by saying, "I studied chemistry and chemical engineering in college. Why am I working in the sales department, marketing department or human resource department?" The supervisor answered, "What you studied in college is what you think you are good at. We want to find out what you really are good at."

The supervisor's wisdom and logic was so overpowering that it overcame what Nadji describes as his young and inexperienced arrogance and helped him realize that what the company was doing was actually good for the company and good for him. During the period when he was transferred to the sales department, on his first day, the sales manager came to Nadji and gave him a tie clip on which the following abbreviations appeared: YCDBSOYA. Nadji asked him what it stood for. The manager said, "It means, You Can't Do Business Sitting On Your Ass."

In other words, the sales manager wanted Nadji to travel company-to-company and sell. However, the telephone along with digital and social marketing have turned historical paradigms upside down. We know that today, an increasing majority of sales take place online or over the phone.

The CEO Had A Good Idea

After the second year of publishing Telemarketing® magazine, one day Larry Kaplan, CEO of Tele Business USA, an outstanding B-to-B teleservices outsourcing company, called him and said, "Nadji, do you know what is the greatest achievement of telemarketing?"

Nadji gave him a half dozen suggestions, but he didn't buy any of them. Then Nadji gave up and asked Larry, "What do you think is the greatest achievement of telemarketing?" He answered, "Telemarketing has proven that YOU CAN DO BUSINESS SITTING ON YOUR ASS!"

And when you think about it, this is 100 percent right. This must have been the greatest evolution in sales: that of transforming the sales process from door-to-door to contacting 35 to 40 customers per day vs. two or three companies or customers per day via the door-to-door method. And, of course, we have learned that selling is a numbers game. He or she who makes the greatest number of contacts to the right audience sells the most.

Jack Welch's Philosophy on Sales Management

A number of years ago, Nadji and his son, Rich Tehrani, had the great honor of having breakfast with Jack Welch, the legendary CEO of General Electric.

Nadji remembers that day as one of the highlights of his life, being so close to such a great man whose accomplishments were far and beyond any other CEO in the world. In response to what is the most effective way to handle your sales managers, Nadji recalls that Jack replied that there are the following four types of sales managers.

1. There is the type of sales manager who upholds company values, i.e., integrity, always doing the right thing for the customer and legally earning a living. If the sales manager upholds the above values and makes the budget numbers, that sales manager is worth his weight in gold.
2. There is the type of sales manager who does not uphold company values and does not meet the numbers. The solution: Get rid of him!
3. There is the type of sales manager who upholds the values, but does not make the numbers. This person deserves a second chance.
4. Then there is the sales manager who does not uphold the values, but makes the numbers. Jack's solution: Get rid of him!

Among other words of wisdom learned from Jack that memorable day were that:

- a company must always be ready to change,

- five-year plans are a waste of time,
- an effective leader must be a coach and a cheerleader and love to win, and
- employees must feel they can reach their dream by working for the company.

Wow, you can really learn a great deal just from those comments. Naturally, his best-selling book called "Jack: Straight from the Gut" is highly recommended. We have enjoyed reading it, and we strongly recommend every business leader read it.

Not only has the evolution from door-to-door to electronic selling changed the methodology of sales, customer service, CRM and the whole customer interaction area, but it also has completely transformed the sales management process. To be sure, a smile and a shoeshine simply does not cut it anymore. Over the years, as electronic selling has increased selling productivity dramatically, by the same token, it has changed the function of sales management, i.e., managing, hiring, compensation, motivation and recognition. Providing guidelines and controls have also changed.

Heretofore, a personal interview would do the job; but today, not only do you need a personal interview, you also need to do a telephone interview. Once hired, train the person extensively about product knowledge, competitive knowledge and industry knowledge, and above all, instill in the salesperson that he or she must under promise and over deliver and always hold the salesperson responsible for upholding company values and meeting the sales objectives.

To effectively manage the modern sales staff, one must continue to train the people, explaining the value of maintaining an up-to-date database. Daily cleaning of the database is vital to success. In addition, the modern salesperson must know that today is the age of relationship selling. And that means you must be of service to your customers at all times and don't just call to ask for the order.

The Essence of Customer Relationship Management Success:

Focus On Relationships; Otherwise There Is No Customer To Manage!

There are many examples of failure with CRM technology and solutions implementations. The problems are usually due to not developing a focused strategy in advance with clearly defined goals for what success will look like. Many business people lack understanding of what CRM means and unreasonable expectations or a lack of planning on the part of the companies that have installed new software, not on the software itself.

If you understand the essence of customer relationship management, you understand that no company can live without it. Companies live or die from repeat business. Without satisfied customers there will be no repeat business, and without repeat business, companies cannot exist.

Understanding this principle fosters the understanding of why, in spite of an alarmingly high failure rate, CRM technology implementations continue to grow and be a priority in worldwide enterprise applications.

Relationship-Building Should Be Priority No. 1

Successful business-building begins with relationship development supported by technology to manage and provide vital information about customers, always realizing that the human factor must be involved in each and every transaction. In plain English, nothing substitutes people-to-people interactive communication, which is the foundation of relationship building.

Avoiding CRM Failure

The following is a brief, step-by-step guide to successfully managing customer relationships and judiciously deploying CRM technologies.

1. First, take a long, hard look at what your company does and how it does it. Concentrate on and define what your core competencies are, what you provide your customers and what your customers want.

2. Next, look at what processes and technology you have in place currently: what are they, how do they perform, what departments or groups are touched by them, and how current or legacy systems can be integrated into any new solution.
3. Define both short-term and long-term goals (with an emphasis on long-term).
4. Come to a consensus from all departments on what they need and expect from a CRM implementation.
5. Reinforce the fact that this will be a companywide change, and get the management of all departments onboard.
6. Evaluate software and services from several different vendors. Look for systems that allow you to impose your business rules upon them rather than ones that force you to adapt to theirs. Examine how the new system will impact your partners, resellers and/or vendors.
7. Negotiate and purchase the system.
8. Train your people on the system.
9. Study feedback from customers and employees and implement valid suggestions.
10. Re-train your people on the system to remind them of the features and value of it, and stay abreast of new functionality as it's added.

How To Avoid the CRM Graveyard

The following are some of the other contributing factors to the high percentage of CRM failures.

1. Lack Of Know-How

   It seems like every time a new concept comes along that appears to be successful many people jump on the bandwagon without really knowing what they are doing. This, by itself, is the biggest contributor to failures.

2. Implementing A CRM Strategy

   There is a tendency to wish for the benefits of CRM while neglecting the principles that make CRM a success. Many companies jump into CRM without adequately strategizing and keeping the entire corporations needs in mind. It is imperative to analyze your customer relationship needs and match system capabilities to those needs. Failure to do this also contributes significantly to failure.

3. Haste And Lack Of Due-Diligence
   To the extent that CRM technology solutions have produced many successful results, many companies hastily, without proper due-diligence, try to undertake a CRM venture to their detriment.

4. Substantial Consolidation And Innovations
   As in any growing business segment, consolidation occurs in technology industries, which can and does often lead to their failure. This factor, plus continuous innovations, can lead to significant end user failures. Lack of continuity at the vendor level certainly increases the probability of CRM failure.

5. Unreasonable Expectations
   Systems and software are often cited as failures simply because someone chose the wrong system for the wrong reason. CRM implementations encompass a host of challenges, including having the appropriate personnel and rules in place, understanding and managing the scope of the project and managing the data strategy. Once again, lack of due-diligence on the part of the implementer is to blame.

6. Lack Of Proper Buying Strategy
   Because CRM encompasses many departments within the corporation, input and requirements from all other divisions are required before deciding on the CRM solution. In short, improper buying strategy will contribute immensely to failure, obviously.

7. The Bottom Line
   If you are not an expert about every aspect of CRM, selection and implementation, get a true CRM professional's advice and don't wing it because it will not work.

Keep these ideas in mind, and you will be well on your way to designing a winning CRM strategy. Never lose sight of the fact it is the human element in the equation that is the foundation of building lasting relationships.

Boost Market Share with Powerful Marketing And Sensible CRM

Leapfrog Your Competition

There has never been a greater opportunity to boost market share, outsell and out market your competition.
A slow economy may not be popular among many, but, for marketers, it offers the greatest opportunity to seize additional market share and leapfrog the competition through powerful advertising and marketing. Economic downturns offer a golden opportunity to position your company for market-share leadership and maximum profitability when the economic turnaround comes.

Everyone becomes extremely lean and mean by downsizing and reducing costs substantially. It's amazing how businesses, when under pressure, can eliminate so many unnecessary expenses they thought they could never live without, but without which they can actually do very well.

For the savvy businessperson, that means a great opportunity to turn your company around and make it much more profitable and grab a far greater market share. If you are not already, then become the new market leader by leapfrogging over the competition.

The Biggest Mistakes of All

In our judgment, two of the biggest blunders made by corporations are the following.

1) The greatest mistake made by downsizing is laying off the core people who are the foundation of your business success.

2) Many ill-advised senior managers also authorize drastic cuts in advertising and marketing. As far as we are concerned, these

people are making the greatest possible mistake and are thereby inflicting the greatest possible damage to their corporations.

Here is why:

When you cut all marketing and advertising budgets, you lose customers. How can you replace the lost customers and still remain in business? This is a very simple principle of business; yet when looking to cut costs, the majority of corporate leaders still make the mistake of eliminating their marketing budgets. In our judgment, this explains why so many companies go under at such times! It's like cutting off your nose to spite your face!

A Blueprint to Gain Market Share

If you study business extensively, you will reach the conclusion that the new leaders in any business usually arise through aggressive marketing and advertising. Our position is that you must advertise aggressively, even in a down economy.

A Few Success Stories

We have studied companies that have been effective in marketing and advertising, and as a result, their market share improved considerably.

Example 1: The Computer Industry

Dell Computer – Dell exhibited best in class for innovation in marketing and advertising in the down economy of the early 2000s. Whether you were an avid television watcher, reader or radio listener, it was impossible during that time to turn on a popular channel on TV, pick up a business magazine or listen to a popular radio station without running across a clever Dell advertisement.

The ads were inundating all the time, so much so that the average person had likely memorized the company's popular tag line:

"Easy as Dell"

You might ask yourself: What did Dell gain by doing this? Dell historically ranked No. 2 for PC sales behind Compaq. But due to its effective positioning, Dell moved solidly up to No. 1. What was even more shocking at the time was that it was rare to see advertisements by Dell's competitors, which left the field wide open to Dell. Meanwhile, its competitors were making the biggest mistake of all: no marketing, no advertising, no positioning, which means the kiss of death.

Example 2: The Auto Industry

BMW – No luxury car manufacturer has as aggressively and cleverly advertised as BMW. You can praise BMW for practically everything for which we praised Dell Computer when it comes to aggressive marketing, advertising and market share improvement. No matter what medium you are looking at or listening to, you are bound to see or hear a BMW advertisement, which is contributing to its domination in the luxury car sector.

Example 3: The Financial Services Industry

Charles Schwab – Charles Schwab has cleverly created a new niche for itself by differentiating from all other financial services companies. It did that by embarking on an aggressive marketing campaign saying that at Charles Schwab the investment banking side of the business does not control what the brokerage side tells the customers. In other words, there is a higher degree of integrity in what Charles Schwab brokers say.

Example 4: The Bedding Industry

Sleepy's – Anyone who lives within the New York metropolitan area has certainly heard or seen a Sleepy's ad online, in the newspaper, on TV or on the radio. They show up quite often, even perhaps more than once a day. Sleepy's seemed to come out of nowhere a few years back to overtake the former No. 1 bedding company, and it is currently in the No. 1 position. Of course, as we've explained, no one cares about the No. 2 company, it's all about being No. 1. You might ask, how did Sleepy's do it? Simply stated, by aggressive marketing, advertising and promotion in all kinds of media 24/7/365.

The Bottom Line

As all of the above examples point out that there is one and only one way to increase your market share and become No. 1 in your field, and that is by aggressively marketing, advertising and dominating all media including online and social channels.

What makes this marketing strategy so critical today is if you adopt it you are practically the only show in town because most of your competitors are staying on the sidelines waiting to go out of business. Great opportunities don't knock several times, usually just once.

Step-By-Step Guidelines

If you are really and truly committed to positioning your company for maximum market share and profitability, here are a few suggested steps for you to take:

1) The Role of Customer Care™:

You must genuinely try to keep most, if not all, of your existing customers through implementation of what we call, Next Level Customer Care™.

See book by same authors, "Taking your Customer Care to the Next Level".

2) The Importance of CRM:

Implementing a truly functional and sensible CRM solution will provide a framework to manage your important customer relationships in sync with your business process.

3) The Case for Marketing Frequency:

Position and differentiate your company 24/7/365 in an advantageous way and remember that aggressive marketing, advertising and promotion are NOT part-time jobs. A true leader doesn't claim leadership for one week, disappear for six weeks, place a couple of ads and then disappear again for six months. Those types of leaders won't be leaders for long. In fact, they will become followers and in some cases go out of business.

4) On Positioning and Differentiation:

Through your clever positioning and differentiation tactics, be very specific in communicating to the marketplace what sets your product or service apart from your competition. This is vitally important because it gives your customers and your prospects a reason to buy from you rather than from your competition.

5) Remember, if you do not position yourself 24/7/365, your competition will position you in the most disadvantageous way.

6) Market Aggressively:

Maintain the most powerful, aggressive marketing campaign that includes a clever marketing strategy with truly effective and targeted advertising. Remember, there is no shortcut to marketing domination, the greatest market share and success.

The above guidelines are a few of the most vital points you need to keep in mind. Focus on them 100 percent and implement them around-the-clock, 365 days a year if you want to gain the lion's share of the market and leapfrog your competition. And remember that this economy is truly on your side to help you gain your dream market share; make the most of it.

A Winning Total Marketing Strategy

If you asked how many companies we know that have a great marketing program, we would have to say, "Very few." But those that do are the only companies that are prospering.

Now, if you asked how many companies we know that have a truly effective total marketing strategy in place and are actually following through with it, I would have to say, "Even fewer!" This is especially true of mid-size businesses.

Truthfully, we are only aware of a handful of companies like Apple and Disney that give the necessary attention, planning and care required to a total marketing strategy.

What We Observe In Entrepreneur-Driven Companies

- Some companies believe in advertising and do nothing else to help their sales, marketing or customer service departments.
- Some companies don't do any advertising, but believe in major event sponsorship.
- Some believe in direct mail only.
- Others believe in word of mouth only.
- Some believe in door-to-door sales, with no lead generation program in place.
- Some consider sales and marketing departments as necessary evils.
- Some believe in exhibiting at trade shows only.
- Some believe in online ads only, such as Facebook
- Some feel that if they come up with a superior product, they don't need to market it.
- And many ignore the value of public relations, or, at best, have an incompetent PR person.

Ironically of those who believe in advertising, exhibiting or having a PR person, most do a very poor job at it!

Put all of the above together and then throw in unstable economic conditions and what do you have? A sure-fire formula for disaster. The results are a huge number of corporate failures, many of which could have been prevented.

Back to the Basics

If we understand that nothing happens until someone sells something, then we are on the right track. The first question to ask is, "Where do the sales come from?" The answer: is from sales leads. If you agree with that, the next logical question is: "Where do the sales leads come from?"

The answer is from a variety of sources, including, but not limited to, the following:

Awareness: The first step toward development of a winning total marketing strategy is awareness. Once you understand that marketing supremacy, and nothing else, spells the difference between success and failure, everything else has a greater chance of falling into place.

Commitment: Awareness should be backed by commitment, particularly on the part of senior management. Once top management is committed to achieving marketing supremacy, the commitment should be filtered down the line so that everyone's support is gained. In doing so, a company can pride itself not only on its product and service quality, but also on its position as a great marketing company.

It All Begins With PR: The most ignored, least understood and least appreciated part of marketing and marketing communication is public relations. By and large, most small and mid-sized companies don't even have a PR department. And even if they do, more often than not, quality is desperately lacking.

"The American Heritage Dictionary, Second College Edition" defines public relations as "the staff employed to promote a favorable relationship with the public." Given this definition, I respectfully suggest that many companies need to carefully evaluate the performance of their PR departments. It has been my experience that many create the reverse impression. A professional PR department, on the other hand, can open many vitally important doors of opportunity through quality editorial exposure.

Social Media: You must have an active online presence across all appropriate digital and social media portals. Your communications must be fresh and relevant. Response to consumer inquiries and feedback must also be timely.

Integration Is the Next Step

Once you have done the above and have positioned your product or service properly, it comes time to develop a super-powerful, benefit-driven sales, marketing and merchandising message that must be delivered in exactly the same manner by every department of the company, including sales, marketing, PR, social media, etc. Now, you need to upgrade the functionality of every department by utilizing state-of-the-art technology.

Database Development Is Paramount

Hopefully, it is clear by now that without a clean, regularly maintained and segmented database, none of the above is feasible. Let us remember that without a solid database, you cannot dream of having an effective sales and marketing department.

Last, But Not Least Is Customer Care™

All businesses need professional Customer Care™ to survive and successfully compete. The reason is simple – when it comes to effective marketing, you must engage your current customers and develop brand ambassadors.

The main objective is to create awareness among senior management that all things being equal, no company can survive the super-competitive business environment without a sound, total marketing program.

The Importance of Innovation

As has been stated, innovation is extremely powerful. Without innovation, there is no point doing any marketing or promotion at all. Today's business world is filled with mediocre copy cats. Don't be one of them – these types of marketers are here today and gone tomorrow. To be effective, you need to figure out what makes your marketing strategy stand above the crowd. You want to give your prospective buyer the confidence to buy from you without any hesitation.

Data Analytics and Modeling – The New Frontier In Marketing Innovation

As global competition intensifies, marketing executives around the world are scratching their heads to find the most effective way to outsell and out-market their competition.

To be sure, the following are facts of life today:

- Sales, marketing and promotion budgets are getting smaller and smaller while top management is demanding more sales.
- Many advertisements are poorly prepared and are often placed in costly media without a clear strategy, thus results generated are disappointing.
- Believe it or not, some companies are still engaged in shotgun marketing.
- Although every business uses the telephone, many of them are not actually aware of its value in Customer Care™ and validating the customer experience.

If you put it all together, you'll see why there's a feeling of chaos in marketing today. There are, however, some straightforward methods to remedy the situation as follows.

Data Analytics and Customer Modeling: The Cutting Edge in Database Marketing

Database marketing can be defined as the mass customization of sales messages to a highly targeted audience. Like anything else, database marketing has gone through the evolution process and finally led astute marketers to a new frontier called profiling and modeling.

Using the services of a company that specializes in modeling of your customers, you can gather profiles along with demographic and psychographic information on your best customers. As such, database specialists can use your customer profiles to develop a model of prospects that are most likely to purchase your products and services.

Once the model is established, the analytics company can then provide you with a significant list extracted from a general database of millions of businesses. Using profile information of your past and present customers, you can now build a database of most likely prospects for your products and services.

Today's sophisticated analytics tools provide you with:

- increased return on your marketing dollars by eliminating suspects and focusing on the real prospects; and
- reduced costs and increasing sales, enabling you to generate enough capital to develop a massive new database of customers that fit your model and are most likely to purchase your products and services.

Once the model has been established, you can now customize your message to this highly targeted audience and substantially increase the results by taking advantage of the powerful synergy that is generated from the combination of database marketing and innovation.

Innovation: The Missing Ingredient in Marketing

It has been said that each person is exposed to more than 1,000 print ads, billboards, TV ads, etc. every day. If that, in fact, is the case (and we have no reason to think otherwise), it becomes clear that many ad dollars are wasted.

Nadji explains further, "Each Christmas, I receive between 300 and 500 Christmas cards. Some of these cards look very expensive and glittery. I would estimate that some of these cards cost as much as $2 to $3. I appreciate every one of the cards I receive; yet, the one card that made the most lasting impression was not the most expensive.

On the contrary, it was the least expensive and most innovative. The card, enclosed in a window envelope, looked like a check. It read, "Pay 365 days of happiness to: Nadji Tehrani." It was endorsed with the sender's name and company. This memorable greeting card may have cost no more than two cents to produce; yet, it made a great impression on me. This story has a moral: You don't need to spend a lot of money to be noticed. If you are innovative, you will be noticed, no matter what.

Here's another example. Eight years ago, TMC received a quality toolbox from a magazine printer. Shortly after that, TMC began to use the printer's services, and, over the years, the company has given them more than $4 million worth of business! What did the toolbox do? It made this particular printer stand above the crowd. TMC staff was more receptive to the printer. The reason it got TMC's business was not the box, but rather the printer's innovative approach. It did something out of the ordinary. It did not get lost in the crowd of a dozen other printers that were soliciting our business. Yes, innovation pays.

Here's To Originals

Put some innovation into every facet of your marketing if you really want to succeed. Above all, be original. In my office, I keep a small poster that reads, "In a world full of copies, here's to originals." All things being equal, the most original innovators get the lion's share of business, while the rest (including the copycats) get the crumbs.

Urge your sales, marketing and customer service staff to be innovative. Come up with original, new ways of doing things that create a win-win situation.

Innovation is the vital ingredient that can improve every aspect of your business. Sales, marketing and customer service through manufacturing, distribution and the service sector can benefit significantly from innovation.

Let's offer incentives for the most suggestions made by the staff and used by the company. Let us all remember that in business, there is no substitute for innovation.

The Marketing Revolution Continues; Are You with It?

If you ignore it, there may be no tomorrow!

The comment by the great French philosopher may not hold true anymore. The comment to which we refer is, "The more things change, the more they stay the same." While this may have held true in many other cases, it is no longer true in the case of modern-day marketing. We can divide marketers into three categories.

Category 1 consists of those who resist change at all costs, and if any of them are still around, they won't be tomorrow.

Category 2 consists of those who attend every conference to learn what is new, and when they go back to the office, they say, "Not now, it is not the time to invest in new sales and marketing technology." This group may be around only a few months while the first group has vanished!

Then there is Category 3, the realistic marketers who believe that:

A Mind Is Like A Parachute: It Only Works When It Is Open.

This message is aimed at the latter group.

Today's fiercely competitive, global environment has no room for Category 1 or for Category 2. These marketers (if you can call them that) will not be around in the foreseeable future. But Category 3 is composed of the fortunate few with vision – those who have changed all along by maintaining a flexible, positive, can-do-attitude toward a new marketing revolution.

The Marketing Revolution

Over the years, we have witnessed many previously unthinkable changes, globally and politically, and marketing has been no exception. Back in the 70s and 80s, who would have thought that the Soviet Union would no longer exist or that the Berlin Wall would come down? Who would have thought that illegal immigration and domestic terrorism would be the defining issues for our country's political elections? As unthinkable as these things were just a few short years ago, the marketing revolution, which still continues to the present day, was also unthinkable.

A Major Transition

We have seen a major transition from:

- Mass marketing to marketing to the individual
- From direct marketing to database marketing
- From creative-driven to response-driven advertising
- From the technology-driven to the market-driven corporate philosophy
- From feature-driven to benefit-driven promotions to selling
- From shotgun marketing to target marketing
- From door-to-door marketing to one-to-one digital marketing
- From wasteful advertising to productive advertising
- From entrepreneurial marketing to realistic marketing
- From fragmented marketing to integrated marketing

The Revolution & How It Came About

As mentioned earlier, over the last decade we have gone from mass marketing to marketing to the individual.

To appreciate this point, let us define one-to-one marketing:

One-to-one marketing is a powerful, money-making tool that is driven by people-to-people, interactive communications.

The only effective way to market to the individual is, therefore, via the telephone and digital channels. How else could you market to the individual than through people-to-people interactive communications?

We see the word interactive mentioned everywhere: interactive media, interactive sales, interactive voice response, etc., all of which are components of integrated marketing.

In fact, the interactive concept has received major coverage in the nation's leading publications, including: Time, The Wall Street Journal, BusinessWeek, Newsweek and Fortune.

Getting back to the marketing evolution and revolution, one learns that the key to survival is NOT having the best product, but more than anything else, having a masterful sales, marketing and customer service organization. We also learned that a masterful sales, marketing and customer service organization results primarily from phone and digital channels. Let us examine why:

1. Marketers have discovered that as great as direct mail once was, it is no longer as effective as it used to be.

2. We learned long ago that door-to-door marketing is a thing of the past, and is completely cost-prohibitive.

3. As we have stated numerous times, advertising has an extremely important, major role to play; but unfortunately, we must add that more than 90 percent of the money spent on advertising is clearly wasted. I would love to elaborate on this topic, except it may get us off track.

   The bottom line: While advertising is the major ingredient of masterful marketing, few people understand it and even fewer people develop effective, response-driven advertising, hence the waste.

4. With all due respect to voice technology and our strong feeling that there is a major role for voice technology to play in today's marketing environment, we feel that its true applications are more abused than used properly.

5. The telephone and online digital channels, therefore, remain as the only viable alternatives to generate sales.

One-to-one marketing through online, phone and social interactions is enjoying its existence and popularity for, among other things, the following reasons:

- It is cost-effective and versatile
- It is measurable, and it works
- Above all, it enhances:
    - cross-selling
    - upselling
    - irate customer handling
    - collections
    - negotiations
    - account management
    - relationship building
    - problem solving
    - conference calling
    - selling
    - market research
    - lead generation and qualification
    - and a myriad of other applications

Needless to say, you cannot achieve any of the above with the same degree of effectiveness (or cost-effectiveness) with any other form of marketing.

The Evolution/Revolution Continues

For those of us who are keeping up with the latest technological integrations, it is simple to envision the enormous forthcoming changes and evolution that lie just ahead. The result of all of this forthcoming change can be translated to mean that when all of these technologies are developed and implemented, business practice will no longer be the same.

- Sales will be affected
- Marketing will be affected
- Sales and marketing support will be affected
- Advertising will be affected
- Direct mail will be affected

- Customer service will be affected
- And yes, everything else in business will be affected.

We recognize that the changes that new technologies are bringing to our businesses will be profound, literally more than a revolution, where your very existence and growth can be limited only by your marketing genius and imagination. But to feed that imagination and marketing genius, you need to know that these technologies really are and what they really mean in terms of impact to your businesses.

We recognize that these technologies are here, and they are much more than just academic curiosities. To protect our businesses' market share and ensure our success, we need to become prepared to adopt these technologies judiciously so that we can compete, and do so effectively and profitably.

The Nine Guidelines for Market Share Supremacy

1. You must think outside the box.
2. You must own online marketing. In other words, you need to be on the first page of leading search engines such as Google, Yahoo, Bing, etc. If it is your goal to be perceived as an industry leader, you must be on the first page of search results. Always remember: If you are not on the first page of search engine results, you don't exist!
3. You must master integrated marketing. There is no other way. You must simply adopt integrated marketing.
4. Master Customer Care™.
5. Master customer service and CRM.
6. Know your competition.
7. Develop a winning technology and marketing strategy.
8. Above all, become a master marketer – remember that the best products without effective marketing are worthless.
9. Partner with the leading media with the largest online reach. To be effective, an online medium must have in excess of one million unique visitors and in excess of 15 million page views per month.

About the Author

Nadji Tehrani

The Man Who Made Teleservices and Call Centers America's Biggest Growth Businesses

Chairman & CEO, Technology Marketing Corporation (TMC)

- Executive Group Publisher, CUSTOMER™ and INTERNET TELEPHONY® Magazines
- Chairman and Founder, Global Call Center Outsourcing Summit™
- Owner of the registered trademark for the term, "Telemarketing®"
- Recipient of the National Leadership Award from the National Republican Congressional Committee
- Honorary Co-Chairman, Business Advisory Council from Connecticut selected by Congressman Tom DeLay, majority leader
- 2003 ATA (American Teleservices Association) Hall of Fame Inductee

"Nothing in this world is as powerful as an idea whose time has come."

Nadji is the founder and Chairman of Technology Marketing Corporation. Nadji is the inventor of the name Telemarketing®. Nadji has trained companies globally by his magazine: Telemarketing® and hundreds of seminars, conferences and trade shows globally.

Nadji has promoted Teleservices companies and ranked them via his development of Top 50 Teleservices companies.

Nadji has promoted Telemarketing® worldwide by conducting conferences and exhibitions. In addition, Nadji has functioned as Marketing Director of Technology Marketing Corporation since 1972 when he founded the company (Technology Marketing Corporation.)

In short, Nadji has been a marketing pioneer and has had over 40 years of marketing pioneering and practical experience.

Nadji Tehrani, by any measurement, is a person driven by powerful ideas. With a deep devotion to originality, he is a visionary who for more than two decades has been able to move powerful new ideas from the laboratory to the marketplace in science, business and industry.

As founder of Technology Marketing Corporation in 1972, he has been publisher of more than a dozen periodicals, books and buyer's guides in the high-tech field, including radiation curing. In so doing, he has established himself as the preeminent spokesman in these important and highly scientific fields.

Since 1982, he has gained even more distinction as the nation's most recognized spokesman for the dynamic teleservices industry. As executive group publisher of CUSTOMER (formerly Telemarketing®) magazine, he is an acknowledged leader in bringing this exciting, multibillion-dollar marketing discipline to the forefront of acceptance in America, as well as other nations around the world. As an industry leader once said, "Nadji has done more for the telemarketing industry than anyone." Among the eminent subscribers of Customer Interaction Solutions magazine, one can find many members of the U.S. Senate and House of Representatives as well as leading universities, such as the Harvard Graduate School of Business.

Telemarketing magazine, which began in 1982, has served as the "bible" in helping companies around the globe, and tens of thousands of people have learned how to increase their sales, deliver superior customer service and build market share like never before possible. The magazine has remained number one in its field since its inception, and is now known as CUSTOMER to better reflect the wide spectrum of businesses it serves, which now includes CRM (customer relationship management), e-sales and e-service.

As a leader in the fast-growing information industry, Nadji Tehrani's educational background is equally impressive. Educated in Europe (Sorbonne: The University of Paris), the Middle East and the United States, Nadji has an undergraduate degree in chemistry and has completed graduate studies in business administration.

Before starting his own company in 1972, Nadji held a number of important research, marketing and management positions at E.I. DuPont, Phillip Morris and Stauffer Chemical.

Although Nadji is multilingual, the language he speaks most fluently is the one that business people all over the world understand-that of increased productivity and enhanced profits through teleservices.

In February 1998, TMC introduced INTERNET TELEPHONY® magazine, the authority on voice, video, fax and data convergence. INTERNET TELEPHONY provides complete coverage of this exciting, emerging technology that will revolutionize communications. The magazine went on to launch its own highly successful show, ITEXPO™, the first of which was held in October 1999.

Nadji is the founder of the Global Call Center Outsourcing Summit™, which is the successor to TBT™ (Telemarketing and Business Communications), the world's original and most comprehensive exhibition and conference for CRM, Internet telephony and contact center industries. Over 150,000 top corporate executives are trained annually through CUSTOMER magazine, INTERNET TELEPHONY magazine, the biannual ITEXPO and the Global Call Center Outsourcing Summit.

Among many of his international achievements, Nadji was once selected as one of America's top 500 corporate executives to visit the Kremlin Palace and exchange views with a distinguished delegation of Russian business executives headed by then-President Mikhail Gorbachev.

In 2003, Nadji Tehrani was inducted to the ATA Hall of Fame. The American Teleservices Association is the only association dedicated to the teleservices industry.

Nadji has spoken worldwide on telemarketing and integrated marketing. Domestically, he has addressed dozens of prestigious organizations, such as Sales & Marketing Executives International, the American Red Cross, the Direct Marketing Association, and the National Convention of the American Telemarketing Association as well as at TMC's own Global Call Center Outsourcing Summit. Internationally, he has spoken in Japan, London, Rotterdam, Paris, Mexico City, Canada, Monte Carlo and Brazil.

Nadji's advice is continually sought by the largest financial institutions on Wall Street and investment bankers for his insight on the industry. He is further consulted by a branch of the Federal Government for assistance in conducting the year 2000 Census through telemarketing. In addition, Nadji has been consulted by the Federal Trade Commission (FTC) on matters pertaining to the telecommunications industry.

Nadji has frequently been interviewed by many prestigious national publications, such as The Wall Street Journal, The New York Times, Chicago Tribune, the Washington Post, Fortune magazine and many other esteemed publications around the country regarding matters concerning the contact center industry, as well as industry statistics. Statistics provided by CUSTOMER™ magazine have frequently appeared in the national press, including the Wall Street Journal, and they are permanent part of the congressional records pertaining to the preparation of the Telephone Consumer Protection Act law.

Technology Marketing Corporation is regarded worldwide as the only credible source of information on the contact center/CRM industry, with over 1,000 audio training tapes, 10 books, six conventions and a 20-year library of information on the industry. As such, Technology Marketing Corporation is the world's leading source of information on the subject of teleservices, customer service, sales and marketing, CRM and Internet telephony. Technology Marketing Corporation owns the registered trademark for Telemarketing®.

Last, but most important, Nadji relates to people. He has dedicated his business and professional career to the enhancement of ideas that enrich the lives of others. Such is the measure of this man and his ideas. Powerful ideas whose times have come…now and for times yet to be!

TMC
CUSTOMER INTER@CTION Solutions
July/August 2012 • Vol. 31/No. 1
www.cismag.com
31 Years
#1 In CRM, Call Centers And Teleservices Since 1982™
The Dawn of The Call Center Industry
& A Brief Biography of The Industry Founder, Nadji Tehrani
Also In This Issue:
• The History and Advancement of the Contact Center and the Customer Experience
• InfoCision Shares 30 Years of Call Center Wisdom
• The Evolution of Self Service

All-New Guide To Contact-Management Software • Preventing Power-Surge Disaster • Facsimile, Supplies & Sources • Marketing To Ethnic Groups & Seniors • Business Applications Of Telemedia — Column Premiere

Telemarketing®:

The Competitive Edge For Every Business In The '90s™

Book Supplement: Telemarketing® Magazine, 35 Years Later, by Founder and Publisher Nadji Tehrani

What a pleasure it has been the last 35 years to be involved with Telemarketing® magazine, which eventually evolved into what today is CUSTOMER magazine.

It seems as if it was yesterday when I was sitting in my office and trying to figure out how I could possibly increase the effectiveness in the sales and marketing area. In 1979, TMC was a fledgling start-up company; we only had four publications in the energy efficient non-polluting and material efficient technologies.

These were as follows:

- Journal of Radiation Curing
- Journal of Waterbourne Coatings
- High Solids Coatings
- Powder Coatings

These were energy-efficient, material-efficient and non-polluting technologies that were very much needed by the industry. As an entrepreneur, I realized that there is, indeed, a tremendous need for such publications.

What was the spark that led you to establish TMC?

Prior to starting TMC in 1972, I had worked with four major companies, three of which were Fortune 500 type companies. The other one was a mid-sized company. At every one of these companies, I looked for the most challenging project, and I approached management to give me those projects and let me come up with a solution for them.

In the summer of my third year of college, I got a job at a company called The United States Bronze Co. located in Flemington, N.J. As the name implies, this company manufactured bronze powders. Upon preparing a batch of 50,000 pounds of the stuff, the company had to wait 24 hours to figure out the percentage of the pigments, namely bronze was composed of 85 percent copper and 15 percent zinc.

This fact bothered me very much, and I went to the president of the company and asked for his permission to try to speed up the evaluation process in the laboratory so that the production could proceed practically upon completion. The president gave me the green light and, to make a long story short, I developed a new electrolyte solution that would speed up the analysis of copper and zinc within the powder in less than 20 minutes.

Obviously, this was a major achievement. The president of the company wanted me to consider becoming laboratory and research director. Since I was unable to accept that because I chose instead to complete my college education, I gratefully acknowledged his interest in me. And I told him that if there were any other major problems, I would like try to solve them as my next challenge.

I wanted to continue working for US Bronze as part of my research project at the college and to try to develop a technique for a satisfactory plasticization of bronze powders into vinyl plastics. To the extent that 85 percent of bronze powder was copper, during the traditional heating and pressurizing, oxidation will occur and, therefore, the bronze pigments would lose their decoration value within the vinyl plastic.

My project, therefore, was to find a solution to prevent the oxidation and allow bronze powders to be used within vinyl plastics. As a result, the company could expand its market share by going to tabletops to furniture-tops of all kinds with a generally white background of vinyl accented by golden sprinkles of bronze powders.

I took this project to the university and inside of 30 days I was able to come up with a new solution to prevent oxidation and keep the glitter of the bronze powders. This discovery was written about in many newspapers around the country, including in Virginia where my college was located.

Shortly, thereafter, I was contacted by the Iranian Embassy in Washington, D.C., and the cultural attaché told me that his majesty, Muhammad Reza

Shah Pahlavi, would like me to join him in his palace in the summer of 1964. I asked the cultural attaché, "Why does his majesty want to see me?" He referred me to my humble discovery at the university. At this point, I was engaged and I asked my fiancé, Julie, to join me. She was very happy, and she was pleased to meet with his majesty, the Shah (or the King) of Iran.

Going back to what sparked me to start TMC, immediately after college I was hired by a Fortune 500 chemical company at which time two of the most difficult projects were given to me and the assignment was to find a solution. To make a very long story short, I was able to invent new processes to eliminate the discoloration of a solvent during the purification process.

Given that the cost of the solvent was $40 per gallon (in 1962), this was a major achievement, and the company gave me a 10 percent increase, which was approximately $60 a month. Then the next project was given to me. In this particular case, the name of the project was waste reclamation process. The company was trying to make polyurethane textile fibers and because of the sensitivity of polyurethane fiber spinning, generally about 30-40 percent of the produced yarn was classified as waste.

Considering that this fiber could be sold at $40 a pound, one can imagine that the amount of waste produced per month could be very close to millions of dollars. Again, this project was given to me with six months of feasibility studies followed by five years of process development.

I was very excited about getting the No. 1 problem of the company, but I didn't want to wait that long to get it done. I went to my supervisor and I said "Len, I don't want to wait 5½ years to solve this problem. I'd like to solve it in one month." He said, "Rookie, I like your enthusiasm; tell me what you need to do to get it done in one month." I told him "I'd like to run three shifts round the clock; this way one month would be equivalent to three months and hopefully by then I would get the project done."

He was slightly hesitant but went along with me anyway. He gave me two additional technicians and, I found that we could add 25 percent of the wasted yarn to fresh yarn. That minimized the amount of wasted yarn and solved another major problem, which was yellowing of the fiber upon washing it in the washing machine.

My achievement saved the company millions of dollars, yet I was only getting 10 percent pay increases.

That was not acceptable to me, and the idea came to mind that if I can achieve so much for other companies and get so little reward, why not start my own company. That served as the spark that gave me the encouragement to start my own company in 1972.

What was the name of the first TMC magazine, and what were the areas of coverage?

Radiation Curing was the first magazine that we started, and it basically dealt with an energy- and material-efficient as well as non-polluting technology. The conventional technology used solvents in preparation of polyeuration products while consuming 100 watts of electricity per pound. But with radiation curing the amount of electricity needed would be about 2 watts and prevent the need for evaporation of the solvent, which meant less pollution.

At that time this type of technology was very much in demand and prompted me to add the other publications, namely The Journal of Radiation Curing, Waterbourne Coatings, Powder Coatings, High Solids Coatings, etc.

What did the call center business look like 35 years ago?

I think if I explain how we came about the idea of Telemarketing® magazine that might help to answer that question very clearly.

In the late 1970s, our publications were flourishing; I noticed that there was a major deficiency in advertising and marketing sales areas. In fact, we had two outside sales reps who were traveling door-to-door trying to sell advertising and submitting significant expense accounts every week, and they were hardly justifying the expense.

One day, the idea came to me that what if I used the phone and tried to sell some advertising? And low and behold, while working one hour a day, I sold three pages of advertising in the first day. I continued this process for the rest of the week, namely spending one hour a day. By the fifth day; I was able to sell 15 pages of advertisements without a car allowance, expense account, food and beverage and the rest of the nonsense that was going on with outside selling. I said, "Wow, this technique can be a new way to sell and do marketing."

Mistake No. 1

I brought in both of the outside reps and put them on the phone, and they quit. At that time I didn't know why, but today I know why. The reason was that the techniques you need to sell in person are completely different from techniques you need to sell on the phone. In person, you need charisma, wining and dining, body language, etc.; whereas on the phone, you have none of that stuff. The only things you need to have are wonderful communication skills, product knowledge and a likeable presentation technique.

Mistake No. 2

I interviewed in person the people who were supposed to sell on the phone. That did not work, and I was beginning to get a bit discouraged because I couldn't be the only person on the planet who could sell on the phone. What a dilemma!

We had a secretary who was with me for five years and knew all about the publications – not necessary scientific questions, but things such as the audience difference, circulation difference, etc.

So I asked that secretary to come to my office and I said, "Geri, I have a $500 reward for you."

She said, "Before you continue, I am the world's worst salesman."

I said, "You didn't let me finish. Everything I have done so far has failed, so you are my only hope. The reason is that you have product knowledge and, more importantly, you have the most pleasant conversation technique on the phone of anyone I know. You have to be able to do it."

Then I said, "We have a $500 reward for you, and just to relieve the tension, you don't have to sell anything."

She said, "I don't have to sell anything?"

I said, "No, all I want you to do is sit down. I'll get a customer on the speaker phone, and you notice how I make the presentation. You just do what I do on the phone, and you will be surprised how successful you will be."

This was in November near the holidays, and she needed the money; therefore, she gave it 100 percent. Low and behold after three months, sales went up by 50 percent with no car allowance, no F&B expense, and no lodging expense.

I thought, "Now we have two people who can sell by phone." I still didn't know why.

Today, I know why. To sell on the phone, you need a pleasant voice, you need to be articulate and you have to have product knowledge, industry knowledge and competitive knowledge.

Having worked five years for TMC, Geri had just about all of the above. And that's why she was successful. Three months later, a trade show came along in Boston, and I took Geri with me to the show.

You would hear the conversations going like, "Nice to meet you." "With your voice, I didn't know how to expect you as a person." At the end of the trade show, within a month, sales went up an additional 60 percent. And the business was rolling. All I had to do was to find other people and train them.

So the first thing I did was file for a trademark registration for the word, Telemarketing. The trademark registration was assigned to me, and I still maintain the trademark.

Shortly thereafter, my editor of Telemarketing® magazine came to me and said, "Mr. Tehrani, are you sure this is an industry? I can't find anything to write about." I said, "I am not sure, but this is called a pioneering project. We are trying to go to uncharted territories."

Then I contacted John Weinman, who was the vice president of long lines at AT&T. I went to his office and showed him the first issue of Telemarketing® magazine and told him that we needed his advice and support if this magazine was to exist.

John said, "If you do a good job, I will support you." Then I showed him a few copies of Radiation Curing and some of the books that we had written at that time.

I said, "John, if we can do these sophisticated chemical publications, I think we can do the telemarketing publications a lot easier."

He agreed and called five of his assistants and asked them all to find telemarketing articles in various companies such as MCI, Uniroyal, and others that were just beginning to try telemarketing. In addition, John brought to my attention that in Chapin, S.C., there was a man named Chilton Ellett who also started a small telemarketing company in his garage.

So I was grateful for John's assistance and upon returning to the office, I called Chilton and asked him if he would like to write an article about his success in telemarketing.

Chilton laughed at me and said, "I didn't finish high school, I don't know how to write and I don't know how to read, but I have a successful tiny telemarketing operation."

I said, "What if I develop a bunch of questions, put them on a cassette, and you play the cassette and then you answer on the cassette and send it back?" He said, "That is easy and I can do that." I said, "Chilton, you are very clever." He said, "Dinosaurs weren't flexible so they vanished; mankind was flexible and we are still here."

Thirty years ago there were only bits and pieces of small call centers scattered around the country.

I later discovered that Omaha had the potential to be the capital of telemarketing. So, I got on the plane and went to Omaha, where I met with Gary and Mary West who were running their business in very humble surroundings similar to an enlarged garage.

Mary took me around and showed me every department, how they worked, what they did, etc. And, by the way, that company grew and expanded and was sold a few years ago for the tidy sum of $2.4 billion.

Across the street from Gary and Mary West's operation in Omaha was another division, the outbound division, which was run by a young aggressive man named Steve Idelman. I spent about a week between these two companies and learned whatever I could.

At that time, the telephone sales reps were sitting about two to three feet from each other and calling people until 11 p.m. I felt that I got quite a bit of information from these two companies which eventually turned out to be the No. 1 and No. 2 companies in this industry in the country.

Steve's company, ITI, was sold for $120 million in the 90s.

I was fortunate enough to learn from such leaders as Gary and Mary West and Steve and Sherri Idelman so that I could build on it. These folks helped me throughout the decades. West Corp. advertised in Telemarketing® and Customer Interaction Solutions® magazine for 20 years with a two-page spread located on the inside front cover and the page across from that.

Steve supported the magazine by placing an ad on the back cover of the publication for several years.

These folks have been extremely loyal. In fact, whenever I traveled to Chicago, Steve and Sherri Idelman would fly from Omaha to Chicago to have dinner with me in their penthouse apartment.

The bottom line is that the call center industry, at that time, was very primitive; namely that there was no technology except for recording of orders, which happened on a tape recorder and basically that was it. There was no other technology.

How did its evolution lead you and the TMC team to evolve the call center/Customer Care™ magazine over time?

Obviously, Telemarketing® magazine was a pioneering publication and in fact I might add that in 1982 we submitted for registered trademark of telemarking, and the U.S. patent office granted me with a trademark for Telemarketing® magazine.

Having said the above, and as the leading publication of the industry, myself and other editors kept up to date with practically every event in call center, telemarketing and customer interaction areas. It was our primary responsibility to share with our readers the new changes and new technologies that were forthcoming and prepare them for being early adopters of these technologies.

Therefore as the industry evolved from telemarketing to call center to customer interaction center and Customer Care™ center, we also developed the magazine to address all of these new changes and so on.

Why was the magazine renamed to CUSTOMER?

In 2002, the industry had evolved to customer interaction and past the outbound only or inbound outbound only, which defined the telemarketing at that time.

As the industry grew, the users found more and more solutions for which telemarketing was used, and customers and vendors began to name themselves as customer interaction technology or service providers. That is when we came up with the name Customer Interaction Solutions. In plain English, it was more appropriate for the leading publication of the industry to stay with the changes in the industry; and therefore we adopted the name Customer Interaction Solutions, which further evolved to CUSTOMER.

What do you see as the prevalent trends in Customer Care™ operations today?

Around the year 2000, an evolution started out in the call center industry, namely that by using internet telephony techniques, countries such as India and the Philippines and others had access to a very cheap long-distance price coupled with very low labor cost, which in some cases was as low as $1 per hour as opposed to $6 to $10 an hour in this country.

That meant that a lot of the companies, which were at the direction of the accounting departments, decided to send practically all of the call center requirements into India the Philippines, Latin America near-shore and off-shore, etc.

At that time, I was uniquely against going to those places particularly in the area of customer service, Customer Care™ and customer interaction. Against all odds, I predicted that these companies within a few years, and after they lost plenty of their customers, would return back to the United States. At that time, everyone thought that I was crazy.

But today, I am happy to say that many of the people who went to such countries as India and the Philippines and others have returned back to the United States for a lot reasons such as:

- Difference in cultures
- The significant time difference between those countries and the United States. For example, day time in the U.S. is equivalent to night time in India and elsewhere; therefore you have TSRs (Teleservice Representatives) who have been working one to two shifts, and particularly those who worked in the night shift were tired and sleepy, and to make matters worse they were treated extremely poorly.
- The TSRs in those countries had to produce by making a significant number of calls and when they left their offices, they had to walk one to three miles to get home. On the other hand, we learned that in India approximately 100,000 people speak proper English so the American public could understand them; whereas, the rest of the TSRs did not speak proper English, and it was difficult to understand them. All in all, the offshore evolution of the call centers turned out to be much more complex than originally anticipated.

Today, the main trend is that Customer Care™ operations are returning to the United States to provide proper service and professional treatment for their U.S. customers.

A great example is the recent merger of Time Warner Cable and Charter Communications. The new combined company is called Spectrum.The focus in its advertising is a renewed emphasis on customer experience and the fact that it plans to bring 10,000 call center jobs back to the US.

I also see that now call centers are beginning to embrace such areas as digital lead generation, and social marketing among many other areas.

What do you see as the prevalent trends in Customer Care™ and customer experience today?

In my judgment, one of the most important parts of the business is customer retention.

To that end, one might say that if you have a bucket of water that has several holes at the bottom, you will never be able to fill it up because as fast as you add water, by the same speed the water goes out of the bottom.

That scenario works well in the business community, namely that if you do not focus on customer retention, customer satisfaction and the overall customer experience, the company will not succeed.

In my humble opinion, companies live or die from repeat business!

Henry Ford once said that, **"A good design sells the car, but quality and performance brings them back."** This concept certainly works in every aspect of business. Customer Care™ and customer retention are all major parts of business.

Another most important part of Customer Care™ is **lead generation**. The sad part of this is that the universities' marketing or sales divisions do not even address lead generation as a viable part of education. **In fact, one might say that companies will grow as much as a high-quality lead generation would permit them to grow.** And in my humble opinion, some of the best kinds of lead generation are through webinars and channels and white papers, all areas in which TMC specializes.

In fact, one might add that the best quality leads are obtained from digital lead generation and/or webinars. Here is a reason why: Suppose your company produces predictive dialers; if you sponsor a webinar under effective selection of a predictive dialer, your quality should be exceptionally strong because the people who sign up for this webinar should be 100 percent interested in predictive dialers or else they will have to see a psychiatrist! Consequently, for high technology products, lead generation via webinars is as good as it gets.

How does that relate to what's happening in marketing overall?

As a student of marketing for nearly 50 years, I have learned a number of things, among them:

- **If you don't market, you don't exist.**
- **If you're not on the first page of Google under your keyword, you don't exist.**
- **If your marketing is not a cutting-edge type, namely, if you're not involved in digital marketing, online marketing, social marketing, then your company will not be going anywhere.**

As far as the impact on marketing from lead generation, the job of the marketing department is to create leads for the sales department to sell to and or convert them to customers. In that case, I see a strong relationship between marketing today and lead generation that goes along with customer retention, customer acquisition, etc.

How is it – and should it – impact decision making by corporate executives?

A number of years ago, I wrote an article titled, **"The Sad State of Marketing in Corporate America"**. The way we came up with that topic was the fact that I had developed a marketing test for candidates who applied for marketing position. I would consider that test to be a simple one; for example, the first question was to define marketing and the second question was to define sales.

Would you believe that 90 percent of the candidates could not provide a clear definition for marketing or sales? I remember in one case a Columbia graduate with a master's degree in marketing and 20 years of experience with a Fortune 500 company could not pass our marketing test.

So I began to wonder, what are they teaching to these marketing students today so that we can hardly find anyone who understands marketing the way they should and remember that if you don't market, you don't exist?

How could these corporations ever survive?

Most of them don't have a clue about marketing. It really doesn't matter what they think about what we have written so far.

Where do you see the call center/Customer Care™ industry going from here?

As I have indicated in many of my past editorials, no company can exist without the telephone. And with the telephone, one would do inbound and outbound calling. I know for a fact that in our company, if you take the phones out, there will be no company.

And, I believe that will continue.

Yes, digital marketing and email have important roles to play. Yes, social marketing has a critical role to play, but when that telephone rings someone has to pick it up. And that could be a prospect or a customer asking for help, and that is where Customer Care™ and customer satisfaction come in.

As for where customer satisfaction and Customer Care™ and the call center industry is going, it is my judgment that all of these industries will continue to flourish as long as our basic business remains the same.

We also know that consumer calling has become a mobile business with landlines playing a minority role in customer communications. And, there are significant government regulations and risk involved in contacting mobile numbers, namely the importance of gaining appropriate prior consent from the customer.

If there's just one thing you've learned over the years about customer interactions that you'd like to share, what is that one thing?

In my judgment, the most important fact that one learns from years of experience is the importance of quality of performance. As I have stated in the past editorials, quality and marketing are not part-time jobs!

I recall in the mid to late 1990s there was a tremendous amount of interest for call centers on Wall Street. Company after company would go public. A given stock, for example, began at $15 per share as an IPO and would go up as high as $80 and split two for one, and the split would go up as high as $80 again.

It was literally a flourishing market for all of the call centers because of the rapid growth, etc. At that time, in the telemarketing tradeshows of ours called TBT (Telemarketing and Business Telecommunications) on many of the keynotes at our shows I reminded the attendees that all good things will come to an end.

And about the only way you can keep this marketing environment is by having an outstanding quality, because quality brings back customers and creates customer satisfaction. Therefore, the industry would be in high demand.

Indeed, many companies listened to my advice and kept the quality extremely high. On the other hand, approximately 30- to 40 percent ignored my comment and shortly thereafter, they vanished. **The most important thing I learned in the last 30 years was that there is no substitute for quality, and quality and marketing are not part-time jobs!**

Today, a lot of advanced technology is available to help call centers vastly upgrade the quality of service.

Farewell and Time for the Changing of the Guard

After many years of industry leadership, in 2009 I decided it was time to ask my son, TMC CEO Rich Tehrani, to begin writing my Publisher's Outlook editorials.

It has been an incredible ride to be at the center of a market that we started, grew and evolved to the point where it is a global multi-trillion dollar market and changed the way all companies do business and allows customers to have closer connections with the companies they purchase from.

It has been tremendous to be involved in taking this industry global, ranking many of the players via a number of awards and, moreover, helping service agencies get on the map and become the heart of a multi-billion dollar outsourcing/BPO market.

How exciting it was to apply for and receive a registered trademark on the term telemarketing back in 1982, when it was barely even beginning. This was five years before the term call center even became popular. Moreover, watching this publication evolve with an industry that changed the world of commerce and communications has been even more exciting and rewarding.

A plaque presented to me at TMC's 25th Anniversary Celebration reads as follows:

25 Years of Excellence
Presented to Mr. Nadji Tehrani
Teleservices Industry Founder, Leader and Visionary
May 4, 2006
"In just 25 years, we went from non-existent to a
proud industry, which laid the foundation for every
corporation in America and around the globe."
"Every company is a call center!"
"Quality and marketing are not part-time jobs."

On behalf of the industry, InfoCision Management Corp.
awarded Nadji Tehrani this prestigious plaque recognizing him
as "Teleservices Industry Founder, Leader and Visionary."

With Blood, Sweat and Tears, We Got This Industry Going

I shared earlier in the Q&A section how TMC championed the telemarketing industry.

To further explain the evolution, let me add a few more details: In the spring of 1983, in this editorial, I called for the automation of the telemarketing industry and, sure enough, many entrepreneurs agreed with me and started developing all kinds of different software and hardware products to automate the handling of calls.

In 1985, we launched TBT (Telemarketing and Business Telecommunications), the world's first call enter event in Atlanta, Ga. To make a long story short, this show was extremely well received and, in fact, conferences sold out and people were actually standing in the hallway listening to the lectures inside the rooms. At that time, I recall receiving a phenomenal testimonial from the president of one of the software manufacturing companies, who wrote me the following testimonial:

> Dear Nadji:
> Congratulations on the success of TBT '86!
>
> Early, Cloud & Company applauds your grasping the leadership baton on behalf of the young telemarketing industry. Early, Cloud & Company will be back for TBT '87 as we had a very successful show.
>
> To date, we have proposed $3.4 million as a direct result of the contracts from TBT '86. As a vendor, the show was significant – not only did we go through 500 brochures in the first hour, but you brought decision makers to Atlanta which we had direct access to.
>
> ECC applauds your leadership, my compliments to you and your staff.
>
> Kindest regards,
>
> John P. Early, *President*
> Early, Cloud & Company

As such, TBT kept growing and growing, and with it the industry kept growing and growing, and our magazine and seminars and conferences continued to grow and grow.

Going Global

Then came the opportunity to go global. I was invited to take the show and the magazine to various countries, among them, Mexico, Brazil, Japan (twice), the Netherlands, Belgium, Canada, London, Hong Kong and France. In fact, our Telemarketing® magazine was translated into Japanese for several years, and into Portuguese for use in Brazil.

The Evolution

TBT eventually evolved into CTI (Computer Telephony Integration) and then to Communication Solutions, which eventually spun off ITEXPO. Thanks to our valuable partners and sponsors, ITEXPO has been a classic success for more than 20 years, with happy attendees and exhibitors, all learning and buying, selling and networking.

ITEXPO has seen exciting collocated events representing some of the most important new areas of technology markets:

- Smart Grid Summit
- 4GWE Conference
- M2M Evolution Conference (Machine to Machine)
- Cloud Communications Summit at ITEXPO
- Virtualization Summit at ITEXPO
- Digium/Asterisk World
- Startup Camp Telephony at ITEXPO
- IoT Evolution Expo
- Real-Time Web Solutions
- All About the API

As such, TMC events continue to grow as the industry's top destination for information and where you can find all of the important companies in a variety of technology markets.

A Few Final Highlights

The Japanese Exposure

As I indicated earlier, I was fortunate enough to be approached by a reputable Japanese publisher who visited us in our offices and asked to translate our magazine into Japanese. I was impressed with the charm of our new partner and agreed to do so. In a matter of weeks, our magazines were going to Japan and then the translated version in Japanese would come back to us. This was an exceptional feeling, as we were producing a publication that Japan, one of the most advanced countries in the world, was interested in translating.

Moreover, I was fortunate enough to be invited to speak repeatedly at related industry events in the country. The country of Japan, with its rich history, was something beyond the scope of my imagination, and it was an honor to have been invited. The conferences themselves went exceptionally well, and I must share with you a funny thing that happened during one of my presentations:

First of all, the Japanese requested that I speak for five minutes and then stop for translation, and then continue to speak again, and so on. During one of these five-minute speeches, I presented a joke to the audience, the majority of whom did not understand English. However, after the translation, the Japanese laughed as hard as possible.

I was pretty shocked by this reaction, as the people at the conference seemed so serious up until that point, and the joke wasn't funny enough to warrant this reaction. I went to the translator later and asked if he modified my joke. He told me he said, "Our speaker has told a joke, and I want everyone to laugh as hard as possible." Anyway, we made our presentations in Tokyo, Kyoto, and other cities, and were invited a couple of years later to come back to do it again.

The Brazilian Experience

At one of our TBT events, I met with a gentleman, believe it or not, by the name of David Letterman. He came and insisted that he take our TBT show to Brazil, and made many promises about how this would be a tremendous opportunity for all of us. I agreed, and we took our show along with three of our top speakers – Steve Riddell, Judy McKee and Robin Richards – and it turned out to be another fantastic success.

During my personal keynote, I talked about the differences between the requirements of inbound and outbound telephone calls. At the end of my speech (after one hour), someone raised his hand and asked: "What do you mean by outbound?" In other words, no one had a clue about this industry. I thought it was partly funny and, on the other hand, I gave him credit for wanting to learn.

The Mexico City Experience

In the early 1990s, as usual, I stayed late in my office to get caught up with work. At about 8 p.m. one Monday evening, the phone rang and a gentleman from Mexico City was calling me. The conversation went something like this: "Senor Tehrani, I want to come to America and spend $1 million to buy telemarketing technology, and I want to build a call center in Mexico City in the next two to three months."

Then I said: "Do you know anything about the industry?" He said: "No." I asked: "How are you going to buy $1 million worth of technology." He answered: "I went to the telephone company in Mexico City and told them what I wanted to do, and they showed me a copy of your magazine with your picture on it and the general manager of the phone company told me 'If you really want to have a successful telemarketing operation, you must go and meet with this man and whatever he says, you do.' Now, you tell me what to buy, and I'll buy it."

Believe it or not, I invited him to attend our shows, I gave him all possible advice, and he ignored 100 percent of everything we told him and bought whatever he felt like. Later, when I was invited by the Mexican Direct

Marketing Association to come and speak to the audience, this same gentleman asked me to visit his call center. I could not believe my eyes. I noticed that everyone had a computer screen and computers on practically every station, but no one was using any of the computers. Instead, they were doing everything manually.

I asked him: "But sir, why are these people working by hand? That is what technology is supposed to do." His answer was: "Senor Tehrani, in Mexico City, it is more important for the bankers and other customers to come and think that we are using automation than actually using the technology."

The Domestic Experiences

Of course, with producing a magazine, running several trade shows a year and traveling around the globe, not to mention traveling to Washington, D.C., to work with senators and the FTC to try to save our industry from harmful legislation, there was somehow time left for me to spend a good deal of time putting the telemarketing service agencies on the map. These folks, like the old shoemaker, were wearing shoes with holes in them while they were trying to market for everyone else. In other words, none of them had any inclination for marketing or advertising or promotion or anything of the kind.

One day, when I was traveling from Chicago to New York, I happened to be sitting next to a senior executive of one of the leading consulting companies. I asked what he did. He said he was a marketing consultant for senior management. He asked me what I did, and I told him that I published Telemarketing® magazine.

Then he asked if that was really working and I said: "You bet." I asked: "Would you recommend it to your customers to use it?"

He said: "No." I said: "Why not?" And he said:

a.) "I don't know anything about it;
b.) it's too new; and
c.) I don't want to jeopardize our relationship with our customers."

Then I asked: "Why can't you use service agencies." And his answer was: "What is that?"

It was then that I realized that the service agencies in America needed a tremendous marketing push. I decided to become the friend of teleservice agencies, which are now known as (teleservices/BPO companies) and came up with ideas such as the Top 50 Inbound Awards, Top 50 Outbound Awards, Rising Stars Awards, as well as MVP Quality Awards, with the main objective of introducing service agencies to corporate America.

Those who won our awards received an emblem from Telemarketing® magazine and used it religiously on their letterheads, online and everywhere else. For all practical purposes, this was the only claim to fame that the award winners had, and one service bureau was honest with me and said: "You know, we get 75 to 80 percent of all of our new business from your Top 50 rankings!"

Among the elite service agencies that I worked with were WATTS Marketing of America. Gary and Mary West were the chief executives of this company, and I was actually trained by Mary West, who spent a week teaching me about every facet of telemarketing, inbound, outbound, customer services, and more. In addition, I was later trained by Steve and Sherri Idelman on outbound technologies and later, we developed a tremendous relationship with a company called InfoCision, founded by Gary and Karen Taylor.

Since the inception or our relationship, I have noticed that InfoCision alone has won more consecutive MVP Quality Awards than any other industry company, and as such, it is indeed, one of the best, if not the best, call center company in the U.S.

The Greatest Joy in My Life

In my judgment, the greatest joy in life is raising a good family, developing a great company, a great team, having many friends, and having a great vision and changing something in the world for the better. In other words, create something that is superior to previous technologies and/ or ways of doing things. Those of you who have followed TMC know that as technology has evolved and media has changed, we have always striven to be a change agent – helping to evangelize the latest technology innovations.

Book Supplement #2 – A Brief Introduction to Marketing Automation

Marketing Automation – What It Is and What It Isn't

By Paula Bernier, Executive Editor, TMC &

Jeff Dworkin, Principal Consultant, Ghostpoint

As often happens with new technologies, marketing automation/technology has become the latest buzzword within the marketing space. Products that were once sold under such names as email marketing, content curation, list management, inbound marketing platforms, and outbound marketing platforms are all being sold as marketing automation.

According to chiefmartec.com, the marketing technology landscape for 2015 included 1,876 vendors represented across 43 categories. That is almost twice the number of vendors that the landscape included in 2014.

So while there is this big bucket of vendors and products that fall into the marketing automation and technology bucket, we are still left with the question of exactly what is marketing automation. With so many players in the space, you can find a plethora of definitions out there.

Marketing automation is the systematic use of technology to track and manage prospects along their journey from initial acquisition by the system until they either become a customer or are ejected from the system as non-viable. This is done by collecting data from a variety of sources (website activity, live-event participation, social posts, etc.) and by automating certain tasks that are designed to cause the prospect to take action. These actions are then used to establish a prospect's level of engagement with a company's marketing programs.

Based on a prospect's level of engagement, that prospect is either ejected from the system, continued to be tracked and nurtured along his or her journey, or turned over to sales as a marketing qualified lead. The goal is to turn the sales team into a closing machine because it is only receiving truly qualified leads from the marketing team.

Features of Marketing Automation

Because marketing automation is still developing, the features that make up a basic marketing automation platform are still up for discussion. This is the short list that I have been using for some time now.

Segmentation and Targeting

Allows marketers to sort, manage and engage prospects as a group. For example, segmentation is the ability to sort prospects according to the products in which they are interested. Targeting enables marketers to then send different messages to each group.

Email Nurturing/Drip Marketing

Once a customer is placed onto a segmented list, the marketer has the ability to schedule a series of communications by whatever medium the customer requests (email, SMS, phone call, etc.) and measure the prospect's level of engagement with those communications. Clicking an "I Want a Demo" link in a communication certainly tells you something about that customer that sales needs to know right away. Clicking on an "unsubscribe" option tells you something completely different about that prospect.

Lead Scoring

Having a way to quantify the actions of a prospect enables finer tuning of the marketing system. Points are accrued to prospects as they interact with the marketing system. Certain activities can be worth lots of points (attending a webinar or downloading a whitepaper) while other activities can be worth fewer points (opening a blog post or visiting a certain web page). Once the total of these points reach a certain threshold (based on an agreement between sales and marketing) the prospect is considered MQL (marketing-qualified lead) and sent over to sales for the close. There can also be trigger events that immediately cause the prospect to become an MQL. Asking for someone in sales to contact them would be a trigger event.

Integration with CRM

This is critical to getting the real value from marketing automation. Once a prospect becomes an MQL, it must be moved to the CRM system, and the sales team needs to be automatically notified. All the information about what the lead did to become an MQL must be transmitted to sales so that sales team members are prepared to engage with the lead. As the sales team further qualifies the lead and eventually closes, the details of the deal need to be transmitted back to the marketing automation system so that marketing activity can be tuned to generate better leads and so that the ROI of marketing activity can be determined.

Marketing Automation vs. Customer Relationship Management

CRM systems have been around for many years. While many of the tasks of a CRM system are similar to those in a marketing automation system (i.e. sending email), there is an important distinction. CRM is really designed to address leads in a one-to-one manner.

One of the major benefits of marketing automation is to be able to manage groups of prospects and to do this with as little human input as possible. For example, many CRM systems can send an email to a group, but a good marketing automation solution sends a series of emails to prospects based on when they joined the group. It can then take follow-up action based on how the recipient interacts with the email.

Marketing Automation is Not Automatic

Setting up an effective marketing automation program requires lots of planning long before the first prospect is ever acquired from the program, converted, and eventually closed. Once the hard work of planning and implementing all of the pieces of a marketing automation program is done, and you can actually turn on the system (and you have done these steps well), that is when tasks that were formerly completed manually can become automatic. And even after the system is running, there is ongoing care, feeding, and monitoring that has to take place.

The goal is to make automatic many of the tasks that used to be done manually by many sales and marketing staff. The system has to scale, while consuming fewer manual resources. This enables marketers to constantly refine and improve the system so that it churns out better qualified leads for sales people, turning them into closing machines.

The Rise of Marketing Automation Software

As business people and individuals who operate in a world in which technology is playing an increasing role, it's easy to get overwhelmed by the onslaught of new skillsets involved for our jobs and new tools we need to master, or at least consider using, to stay competitive.

A recent issue of CUSTOMER contained a cover story about the rise of marketing automation software, and how that's changing the marketing arena as we know it – by introducing an unprecedented amount of technology and analysis involved. It went on to note that martech pundit Scott Brinker says it's difficult to find this breed of individuals, which he refers to as unicorns and says: "Good luck finding such mythical creatures."

Of course, we are the ones we've been waiting for. The bottom line is that when new tools and trends arise, we often just have to figure things out as we go. The good news is that the process for doing that in this age of accelerating change and new technologies may not be so different from how we've always done it. And sometimes it helps just to know we're not alone as we grapple to get a handle on our new reality.

"At the end of the day, we are all stumbling our way into this," Corey Craig, experience design and innovation lead at Dell, said about always-on marketing and customer experience design, during a presentation recently.

Craig then went on to detail her experience as a customer experience designer and the processes she uses to help in her day-to-day work and the strategy to reach customers. These practices have helped yield Dell, which just two years ago had a 1 percent email click-through rate, reach a 30 percent open rate on its communications with customers and prospects, and three times higher average order values.

Dell wants to ensure that it is where customers are, she said. Customers spend 70 percent of their customer journey online researching products and services before they even talk to a sales rep, she noted, so the purchase funnel is no longer linear. Dell's goal, she added, is to help its customers through all those twists and turns. She added that it's true that "80 percent of success is just showing up."

Let's say a product manager Jim, has a security breach, she said. He fixes it, then he gets online to start researching how to avoid security breaches of his company in the future. So he visits the Dell website. That's an indicator of intent, said Craig.

Content is what nurtures the intent, she added, so Dell makes sure it's at the ready with a targeted email to send to Jim, and an automated system in place to find that email and get it to Jim quickly. Knowing what to send and when to send it is really dependent on your logic – which is where data and content come together, she said, adding this is your "if, then" statement.

While this ultimately became automated, it relies on a lot of manual work to figure out what can and should be automated, she added. Dell has 45 emails ready to go, but deciding on the subject matter of those emails, and then creating and editing them involved a lot of manual effort, she noted.

Figuring out where to start in such marketing efforts, including understanding prevalent customer needs and how best to address them with content, also involves a lot of initial analysis and planning, she indicated. To do that, Craig said, it is helpful to visualize complex ideas.

"We all need to be really great explainers," she said.

For Craig, visualization involved turning in her notebook for a sketchpad and printing up six-foot posters that her team can look at and write on to figure out goals and processes for meeting those goals. The team also creates look books of everything it is trying to accomplish with its program around marketing automation, and it puts those books online so people within the group and from other Dell divisions can educate themselves on the efforts and messaging.

Because so much is involved in marketing automation, she added, it makes sense to start small and start manual. Otherwise, she said, the task can become overwhelming.

And at the beginning of and throughout the process, she added, understand and keep in mind the experience you want to provide Jim and your other customers.

Printed in the United States
By Bookmasters